James F. Haw

**In-Situ Spectroscopy
in Heterogeneous Catalysis**

In-Situ Spectroscopy in Heterogeneous Catalysis

Edited by
James F. Haw

Editor

Prof. Dr. James F. Haw
Loker Hydrocarbon Research Institute
and Department of Chemistry
University of Southern California
University Park
Los Angeles
CA 90089-1661
USA

Library of Congress Card No.:
applied for

British Library Cataloguing-in-Publication Data:
A catalogue record for this book is available from
the British Library.

**Die Deutsche Bibliothek – CIP Cataloguing-
in-Publication-Data**
A catalogue record for this publication is
available from Die Deutsche Bibliothek

© Wiley VCH Verlag GmbH, Weinheim, 2002

Printed on acid-free paper.

Composition pagina media gmbh, Hemsbach
Printing Strauss Offsetdruck, Mörlenbach
Bookbinding Großbuchbinderei J. Schäffer
GmbH & Co. KG, Grünstadt

ISBN 3-527-30248-4

Cover Shown on the cover is the
methanol-to-olefin catalyst HSAPO-34
containing reactants, intermediates,
products, and a deactivating species,
pyrene.
The image is deliberately blurred for
artistic effect but also to suggest that the
mechanism is just coming into focus
through in situ studies.

ac

QD
95
I4
2002
CHEM

Contents

1
Overview of In Situ Methods in Catalysis

James F. Haw

1.1
Introduction

A chemist cannot but wonder at the mechanisms of chemical reactions and how to improve or apply a reaction based on an understanding of mechanism, but this natural curiosity was easily frustrated when first directed toward heterogeneous catalysis. Reactions on the surfaces of complex materials are less easily studied than reactions in solution catalyzed by well-defined molecular species. Of course, homogeneous catalysis and much else of chemistry would be empirical and disorganized without the aggregate contributions of many spectroscopic experiments and theoretical studies. For heterogeneous catalysis, the corresponding investigations are still in their infancy, but we should expect rapid progress as the organizing principles reveal themselves.

In situ studies of reactions on solid catalysts are primarily curiosity driven, and they should be judged by their contributions to basic science. The demystification of any important area of chemistry is an intellectual advance. However, it is also possible that in situ studies could contribute to the development of improved catalysts and processes. The chemical and petroleum industries are catalysis industries. The smallest technical advantage can lead to economic success in these small-margin businesses, and finding a catalyst with higher activity, better selectivity, or a longer lifetime is one of the more common means to a technical edge. Progress in heterogeneous catalysis will be increasingly driven by environmental and resource issues. Natural gas and gas hydrates are abundant hydrocarbon reserves, but the direct conversion of methane to higher hydrocarbons remains a scientific and engineering challenge. The conversion of methane to methanol is an established if only partially understood catalytic process (and it is one of the applications of in situ EXAFS in the chapter by Meitzner). Methanol, in turn, can be used to synthesize olefins or other hydrocarbons on microporous solid acid catalysts, thus providing routes for converting an underutilized greenhouse gas into plastics, synthetic fuels, or other consumer items. The chemical industry is moving beyond pilot plant studies to the construction of integrated units for the conversion of natural gas to polyolefins or other easily transported products. Several of the chapters in this book use studies of methanol conversion catalysis as an extensive example of their respective in situ methods.

Catalyst discovery has usually been an Edisonian process; the rational design of improved solid catalysts is only occasionally demonstrated, often by converting a proven homogeneous catalyst to solid form. Attempts at rational design sometimes falter due to an imperfect or totally inaccurate understanding of mechanism, and methanol conversion is an excellent example of this. Twenty years of speculation, product studies, and analogies to solution chemistry produced at least twenty distinct mechanisms for this chemistry (some with variations) [1]. Many of these required highly energetic intermediates that might be favored by a more strongly acidic catalyst. Recently, it has been realized that methanol is converted to hydrocarbons on solid acids by first methylating cyclic organic species, which may be aromatic hydrocarbons or other species in equilibrium with very stable carbenium ions. These large intermediate species, with typically seven to thirteen or more carbons, act as scaffolds on which bond-making, rearrangement, and bond-breaking steps can occur without the need for very high energy intermediates. In situ studies have made several contributions to the elaboration of this "hydrocarbon pool" mechanism, and several patents recently filed in the area of methanol conversion catalysis were motivated in part by in situ studies of reaction mechanism. Ironically, the microporous solid acid that is emerging as a leading candidate for commercialization is a weaker acid than the one originally used, and its advantages relate in part to the way in which it manages the hydrocarbon pool. In situ studies will lead to major revisions in the mechanisms of other important catalytic processes, and improved mechanistic understanding will make the rational design of improved catalysts more feasible.

1.2
Catalytic Materials

Solid catalysts come in many forms; Fig. 1.1 illustrates three of the more general types. A view down the straight channel of the aluminosilicate zeolite HZSM-5 (also called H-MFI because MFI is its three-letter topology code) is shown in Fig. 1.1 (a). This zeolite was used in the first generation of methanol conversion catalysts, and it figures in some of the applications of vibrational spectroscopy in the chapters by Howe and Stair. Nicholas has done much of his theoretical modeling of zeolite catalysis using a cluster model excised from the HZSM-5 structure. There are over 135 known zeolite topologies [2] and these crystalline oxides can be constructed with some or all of the silicon and aluminum atoms replaced by a variety of other elements. Many of these materials can serve as solid acid catalysts and several can be formulated as useful basic catalysts. Reactions on a silico-aluminophosphate solid acid with the CHA (chabazite) topology are a major focus of the in situ NMR chapter. Other variations on zeolites and related materials are obtained by introducing transition metals into the cages or channels to form high surface area materials with highly dispersed metal components. The chapter by Meitzner describes the use of in situ EXAFS to characterize molybdenum dimers in zeolite HZSM-5.

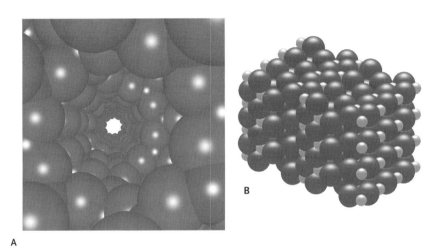

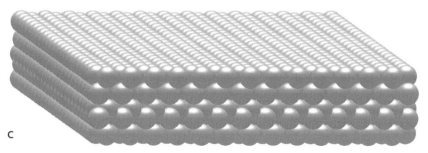

Fig. 1.1. Three of the great many types of catalytic surfaces. **A.** A view down the straight channel of zeolite HZSM-5. **B.** A small crystallite of TiO_2 (anatase). **C.** A low-index plane of the Pt surface.

Figure 1.1 (b) shows a small crystallite of the anatase form of TiO_2. The in situ NMR chapter describes Raftery's photocatalysis experiments [3], in which NMR is used to observe reactions on small particles of TiO_2 coated onto tiny optical fibers. Metal oxides find use as catalysts and as supports for other catalytic components. Anderson and coworkers describe the application of Positron Emission Tomography (PET) to study transport and adsorption on an automotive exhaust catalyst prepared by supporting ceria-promoted Pt on γ-alumina. Mixed metal oxide catalysts are sometimes formulated with two to four components; Hinrichsen et al. describe TAP (Temporal Analysis of Products) reactor studies of propane ammoxidation catalysts containing mixed oxides of vanadium and antimony supported on alumina.

Over half of the transition metals are used as catalysts in their metallic forms, with small amounts of promoters, or in combination with other metals. Metal-based catalysts can be thin films, nanometer-size particles on reactive or inert oxide supports, or simply chunks of metal with or without advantageous impurities. Figure 1.1 (c) shows a view of one of the low-index planes of the platinum surface. Chen and Somorjai describe the application of two methods, sum-frequency generation (SFG) and high-pressure scanning tunneling microscopy (STM) to low-index Pt surfaces.

Whatever the initial structure of a catalyst, it will frequently change under reaction conditions, and it may not even be a true catalyst until subjected to reaction conditions – a caveat often reflected in patent language. The active site of the catalyst could be a well-defined structure present in high concentration, a less-well defined continuum of defect structures, or it may even be something that is created by the adsorption of a reactant or intermediate and that varies in structure over the catalytic cycle. Organic reactions catalyzed by surfaces range from those such as cracking on solid acids [4], which break larger molecules down, to those such as olefin polymerization on supported metallocenes [5], which link smaller molecules. They also include reactions such as selective oxidations [6], which functionalize molecules. Most industrial catalytic processes use high temperatures and short contact times, and many use high pressures. Process design is key to many catalytic processes and fluidized beds and other elegant reactor designs are common. Most processes using heterogeneous catalysis are thermal, but electrochemical catalysis is becoming increasingly important, one emerging area being fuel cell catalysis, while photochemical catalysis has potential environmental and other applications. Clearly, one of the obstacles to a deeper understanding of catalysis is the sheer variety of catalytic materials, reactions, and reactors.

1.3
Compromises

A perfect in situ experiment would look inside an industrial reactor and reveal the most intimate details of a surface chemical reaction, including transition states for reactions on rare defect sites. This perfect sensor would profile chemical composition on the scale of catalyst particles as well as spatial variations over the volume of the reactor. It would have a very fine time resolution, enabling it to follow the temporal evolution of all spatial and compositional variables. Finally, the perfect method would correlate the composition of the catalyst bed with simultaneous measurements of product distributions in order to reveal structure-activity and structure-selectivity relationships. This perfect in situ method is and will remain a fantasy, but individual in situ methods, playing on their strengths, can do a reasonable job of approximating one or a few characteristics of the hypothetical perfect method. Concessions in time scale, temperature, pressure, and/or reactor design must invariably be made for the benefit of the spectroscopic method, and essentially all in situ studies of heterogeneous catalysis are only approximately in situ. Too much emphasis on duplicating industrial conditions can result in an experiment that is so degraded that more can be learned by careful analysis of samples removed from the reactor, and we should take a more inclusive view of what qualifies as an in situ experiment. We are interested in learning more about catalysis, and the real issue in evaluating a spectroscopic study is not whether the measurement conditions meet the purest definition of "in situ", but whether it teaches us about the chemistry that occurs under reaction conditions. The restrictive definition of in situ requires that the measurement be made under reaction conditions, while the inclusive definition allows results that are relevant to reaction conditions.

The conflicting requirements of in situ spectroscopy and industrial catalytic processes lead to compromises in reactor design, catalyst composition, temperature, and pressure. One doubly unique example of an in situ measurement technique creating a compromise in the modeling of a catalytic process is provided by Stair's UV Raman experiment. Stair found that the incident laser beam could cause thermal and photochemical reactions on the catalyst, a uniquely direct case of the spectroscopy modifying the catalysis. The second unique feature of this example was the compromise Stair made in the in situ experiment to solve the problem; he used a fluidized bed to circulate catalyst particles through the beam so as to minimize beam damage. This compromise in reaction conditions for the sake of spectroscopy actually increased the fidelity of the experiment to the industrial catalytic process.

1.3.1
Reactor Design

A laboratory in situ measurement necessarily alters reactor design from that used in an industrial process, sometimes radically. Many in situ experiments make no attempt to simulate reactor conditions. Batch reactors are not commonly used in industrial applications of heterogeneous catalysis, but sealed tubes or cells containing samples of catalyst and reactants are easily handled using standard NMR, IR, or other spectroscopic equipment. Once a batch reactor is sealed there is no further addition of reactant or removal of product. Batch conditions have been used in the majority of in situ NMR experiments to date, and they are sometimes used in most other forms of in situ spectroscopy. Most of the newer techniques for in situ NMR are intended to provide gas flow over the catalyst bed, introduction of reactants, and removal of products. These experiments include high-temperature magic-angle spinning probes with some degree of reagent flow, as well as quench experiments. In the latter case, a fairly uncompromising catalysis experiment in a bench-top flow reactor ends abruptly with a rapid thermal quench, allowing for an uncompromising NMR measurement at a later time. The quench NMR experiment as well as several others in this book raise the question of what constitutes a "true" in situ experiment. Must the spectroscopic observation be made while the reaction is in progress at high temperature, or can catalysis and spectroscopy be separated in time and place? The quench experiment compromises on this point so as to be less compromising in terms of reactor design and other experimental conditions.

Howe's chapter surveys a number of methods of in situ infrared spectroscopy, and in each of them different compromises in reactor design are made for the sake of spectroscopy. Usually, the optical path length through the catalyst must be kept short, which is achieved by using thin wafers of catalyst studied in transmission mode or cups of catalyst powder observed using diffuse reflectance. Many forms of in situ IR permit continuous or pulsed reagent introduction and product removal.

Most forms of in situ spectroscopy treat the catalyst bed as a more or less homogeneous object, and seek to determine the structures of adsorbed reactants, intermediates, and products in order to deduce the reaction mechanism. In real catalytic reactors, reactants, intermediates, and products have spatial distributions. On a stationary catalyst bed, both transverse and longitudinal profiles will develop, and these can be key to understanding an industrial catalytic process. Anderson and co-workers have made considerable progress in the use of PET to determine compositional variations across the dimensions of catalytic flow reactors. PET currently has an effective spatial resolution of several mm, but this could perhaps be reduced by an order of magnitude. Analytical techniques that provide spatial resolution must sometimes sacrifice temporal resolution, and the number of decay events required for a PET scan of a catalyst bed dictates a measurement time of up to 15 minutes. Anderson et al. have made a number of demanding measurements of the diffusivity and adsorption behavior of alkanes under steady-state process conditions. These and similar PET measurements require the use of ^{11}C-labeled chemicals. This isotope has a half-life of 20.4 minutes; its production requires an associated cyclotron and fast, highly selective chemistry is required to convert the initial chemical form of the isotope (typically $^{11}CO_2$) to labeled CO, hexane, or some other reagent that can then be introduced into the reactor. Other methods potentially capable of spatially profiling catalytic reactors are magnetic resonance imaging (MRI) and computer-assisted X-ray tomography (CAT scanning). Neither technique has been developed as an in situ method in catalysis to anywhere near the degree that Anderson et al. demonstrate for PET.

In the case of the TAP reactor system, the reactor design is the essence of the experiment. A conventional bench-top research microreactor provides a time resolution of at best a few hundred milliseconds. The TAP reactor permits experimentation in a regime that lies somewhere between microreactors and molecular beam experiments probing reactions on surfaces in vacuum systems. The design of the TAP reactor permits transient experiments in the millisecond time regime. In this case, the reactor design is not a compromise but a strength, and the chapter by Hinrichsen et al. shows that the TAP reactor can provide information as diverse as diffusivities, heats of adsorption, and catalytic reaction pathways.

1.3.2
Catalyst Composition and Feed

Approximations to the compositions of industrial catalysts are commonly made to simplify in situ experiments. The catalyst may be approximated in a more or less deliberate way, and the approximation may be close or not close to the industrial material. At one extreme, a surface scientist may look at chemisorption and reaction on a series of low-index metal surfaces to understand reactivity on well-defined surface structures. While a single low-index surface is a poor approximation of the metal component of a real catalyst, a series of investigations surveying a number of ideal surfaces and introducing point and line defects builds an approximation to a real catalyst in a systematic way. In some cases, it is possible to grow ordered metal oxide surfaces on metal substrates [7]. There are, however, no two-dimensional models of zeolite catalysts, and zeolites must serve as their own approximations. Binders and other additives may be omitted, the acid site concentration might be increased to facilitate spectroscopic detection of adsorbate complexes, the crystallinity might be increased to improve the resolution of spectroscopic or diffraction measurements, or other approximations may be introduced to favor the experiment. The catalyst may even dictate the choice of in situ method, as some forms of spectroscopy are applicable to some but not all materials. NMR is a wonderful technique for the study of adsorbates on high surface area diamagnetic materials, but strongly paramagnetic metal oxides (not to mention ferromagnetic materials) or catalysts with very low surface areas are more challenging. The most generally applicable experimental in situ method may be infrared spectroscopy, and Howe illustrates in situ IR with applications involving zeolites, several paramagnetic catalysts, as well as both supported and single-crystal metals.

The feed streams used in laboratory in situ experiments may differ from those used in the corresponding industrial processes. A laboratory in situ experiment can use pure analytical reagents to simplify interpretation, but the possible effects of impurities on catalyst activity and deactivation must sometimes be considered as well. Isotopically labeled chemicals are commonly used in NMR, PET, TAP, and IR studies. Kinetic isotope effects can usually be neglected at the high temperatures used in catalysis. The most common compromises involving feed are related to the method of introduction, for example, pulsed instead of continuous introduction or unrealistic space velocities.

1.3.3
Temperature

In situ experiments commonly make compromises in pressure, temperature, or time scale to facilitate the spectroscopic observation. The most familiar example is the pressure gap between real catalysts (typically atmospheres or tens of atmospheres) and the demands of ultra-high-vacuum (UHV) surface probes. The chapter by Chen and Somorjai describes two very promising strategies for bridging this gap, SFG and high-pressure STM. Sum frequency generation probes the surface using two laser beams, one at a fixed wavelength in the visible region, and the second a tunable laser in the infrared. These beams can pass through a high-pressure gas atmosphere with little or no attenuation. One of the unique characteristics of the surface is that it has a very low symmetry, while the gas phase and typically the bulk solid are both centrosymmetric. If the two beams intersect at the surface, non-linear optical effects can result, including the generation of a third beam at the frequency sum of the two input beams. This third beam contains information about the vibrational spectra of surface species only. In STM, the probe tip is typically within a few molecular diameters of the surface, and the presence of gas, even at high temperatures has no bearing on the tunneling current. SFG and STM measurements constitute two bridges between the well-defined two-dimensional materials of UHV surface science and the high pressure world of real catalysis.

The "pressure gap" between UHV studies and real-world catalysis is ten or more orders of magnitude, but smaller pressure gaps, one or two orders of magnitude are commonly seen in NMR and other in situ studies. For example, in 1992, my group reported an in situ NMR study of methanol synthesis (from CO, CO_2, and H_2) on $Cu/ZnO_2/Al_2O_3$ catalyst [8]. The catalyst composition was varied in a systematic way, and various possible intermediates were directly adsorbed and studied. Reactions on the catalyst were observed as they occurred in the NMR probe at temperatures as high as 523 K. However, high pressures are required for methanol synthesis to be thermodynamically favorable, and a typical methanol reactor runs at 50 atm. Our in situ NMR experiments, performed in sealed rotors, invariably had reagent partial pressures well below 1 atm., even at the highest temperature studied.

In situ experiments of all types are sometimes performed at temperatures below those used industrially to relax an instrumental constraint or to reduce reaction rates (lengthen the time scale) for the benefit of spectral acquisition. High temperatures degrade NMR sensitivity by reducing population differences across transitions and increasing thermal noise in the receiver coil. This, in turn, requires longer spectral acquisition and hence a degradation in time resolution. Most in situ NMR experiments require magic-angle spinning (MAS) to average the orientation dependence of the chemical shift, and it is difficult to fabricate a reliable MAS probe that operates much above 523 K. Hot catalysts also emit in the far-infrared, but this is usually not too much of a problem for in situ IR as most catalysts are opaque below ca. 900 cm^{-1}. UV Raman does not have this problem at all.

Temperature and time scale are often inversely related, and some less sensitive in situ experiments are better suited to measurements under steady-state as opposed to transient conditions. Several of the in situ methods described in this book, including the TAP reactor and the quench NMR experiment, were designed with a view towards short time scales and high temperatures.

1.3.4
Theoretical Calculations – A Promising Future

The theorist modeling catalytic reactions must also make approximations to the catalyst structure, and these can be extraordinarily severe. It is common to use a cluster model with fifteen or fewer heavy atoms (i.e., atoms other than hydrogen) to model a catalyst acid site in a zeolite with a unit cell of several hundred heavy atoms. Even so, this cluster model approach has led to some important insights into catalysis, as is illustrated in the chapter by Nicholas. If any technique has the potential to approach the dream of a perfect in situ method, it must be theoretical modeling. In order to truly understand activity and selectivity in a catalytic process, one would need to know the structures and energies of all of the transition states connecting reactants, intermediates, and possible products. No experimental method can provide such information.

A few years ago, it was common to underestimate computational chemistry as a tool for understanding catalysis, but this is changing as a result of less severe approximations made possible through improved methodologies and larger, faster computers. In a few more years, we will see fully periodic theoretical modeling of adsorption in zeolites, which will provide structures, energies, and spectroscopic observables so accurate that discrepancies between theory and experiment could fall either way. Theoretical calculations are inherently more difficult when they must treat unpaired electrons or relativistic effects, but these are just details. A more fundamental problem for computational studies of catalysis in the long run is an insufficient understanding of what surfaces or structures to model theoretically. Experiment will always be needed to point the way, validate methodology, verify predictions, and provide new problems in need of modeling. Papers combining theory and experiment in heterogeneous catalysis are becoming common [9], and they will become more so as the accuracy and availability of computational methods increases.

Experimentalists who are not familiar with theory do not appreciate the extent to which a computational chemist does experiments in much the same way as a spectroscopist. Theoretical modeling does not happen automatically through a completely deterministic process any more than a Raman, IR, or STM study of a catalytic reaction happens without the intellectual involvement of the scientist. Computational chemistry software optimizes geometries yielding energies and force constant matrixes, but it says nothing about the relevance of those numbers or what calculations to do next.

1.4
Spectators

Common to most forms of in situ spectroscopy is the problem of "spectator" species. A spectator is something that forms on the catalyst under reaction conditions and is readily detected by an in situ experiment, but it does not lie on a direct pathway between reactants and products. A spectator is greeted first with undue enthusiasm, falls into disrepute upon scrutiny, and finally is rehabilitated as part of a larger story. The most familiar spectator is the ethylidyne species observed by Somorjai during ethylene hydrogenation on platinum metal. The NMR chapter reviews my observation of trimethyloxonium ion formed in the disproportionation of dimethyl ether on zeolite HZSM-5; this proved to be a spectator unrelated to hydrocarbon synthesis. Spectators remind us that the first unusual species detected on a catalyst is not necessarily the key to the reaction mechanism.

1.5
Future Prospects for In Situ Studies of Catalysis

Our continuing reliance on fossil fuels for power generation, transportation, and consumer goods comes with a myriad of resource and environmental problems, and the management of these problems will require (among other things) a significant, ongoing investment in catalysis [10]. We need better metal oxide fuel cell catalysts, improved automotive emission catalysts, more selective catalysts for partial oxidation of hydrocarbons, and new catalysts and catalytic processes that directly or ultimately use methane as a feedstock. Chemical processes must operate at higher yields, offer greater selectivity, use less energy, and produce less waste. While some of the challenges ahead will no doubt be met through incremental improvements of existing scientific knowledge and engineering practice, some challenges will eventually become so great that only revolutionary advances will suffice. Incremental advances can move forward in an Edisonian way, but revolutionary advances will require that we understand catalysis better than we do today. It will require talented young scientists from varied disciplines.

The future of the in situ approach goes well beyond improvements in instrumentation or new applications. Imagine an in situ experiment evolving into a device for synthesizing new catalysts "in situ", while measuring structure, activity, and selectivity at each step. If one truly knew the structure and function of the catalyst under reaction conditions, and one could manipulate composition and structure within the reactor, one could create new functionalized catalysts while monitoring reaction progress by spectroscopy at each step. Somorjai has taken some steps in this direction by using electron beam lithography to fashion novel surface structures for the study of catalysis by well-defined interfaces [11]. The structures of these surfaces and their catalytic properties are determined within the same apparatus as used for their fabrication. An analogous strategy could be applied, less perfectly perhaps, to real catalysts. For example, microporous catalysts could be modified and tested in a reactor that is also monitored using an in situ form of spectroscopy to guide the functionalization chemistry. Since microporous materials already have periodic structures (channels and cages) on the nanometer scale, reagents metered into the flow stream will form reaction products with spatial periodicities patterned by the host material. In situ observation would guide this process and greatly speed up the synthesis and testing of a catalyst with a specified modification.

This vision is not unique to catalysis; other chemists wish to achieve a more exact control of the structure of matter for engineering applications such as electronics, optics, and computing. Improved catalysts with novel structures should be one of the goals of nanoscience.

1.6
My Introduction to In Situ Studies of Catalysis

The intent of this book is to introduce young scientists, with training in diverse areas, from atomic force microscopy to optical spectroscopy or computational chemistry, to the opportunities for applying their talents to basic research in heterogeneous catalysis. I came to work in catalysis by way of in situ spectroscopy. In the mid-1980s, I was an assistant professor working in the area of solid-state NMR, and my group was using variable-temperature MAS NMR to study inorganic compounds with temperature-dependent magnetic properties. Jack Lunsford, one of my senior colleagues at Texas A&M, proposed an NMR study of propene on an

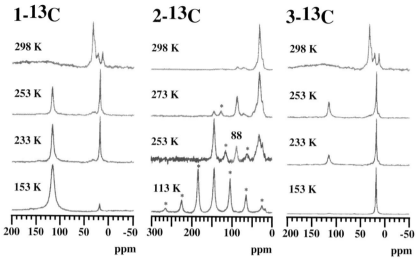

Fig. 1.2. Three in situ ^{13}C NMR studies of the reactions of propene on the acidic zeolite HZSM-5. In each case, propene labeled with ^{13}C in one position was adsorbed onto the catalyst at low temperature, and spectra were measured as the sample was warmed in the MAS NMR probe. These results are consistent with the exchange process shown in Scheme 1.1; * denotes spinning side bands.

acidic zeolite. At that time, zeolites were widely supposed to be superacids, and propene was expected to form the isopropyl carbenium ion (CH$_3$–CH$^+$–CH$_3$) as a stable species upon adsorbtion into a zeolite. We adsorbed propene on the zeolite at room temperature, observed a ^{13}C resonance at $\delta = 250$, and assigned this signal to the isopropyl cation [12]. In superacid solution, this signal is observed at $\delta = 320$, and we assumed that weak solvation by the zeolite accounted for the attenuated value.

A little later, my research group was using solid-state NMR to look at chemical reactions within MAS rotors, and a follow-up on the propene study was an obvious application. We adsorbed propene onto an acidic zeolite at cryogenic temperature and transferred the sample to an NMR probe without heating. By gradually raising the probe temperature and acquiring high-resolution ^{13}C spectra, we expected to see first unreacted propene and then the formation and reaction of the isopropyl carbenium ion. We saw the former but not the latter. The signal at $\delta = 250$ was the last to appear, near room temperature, at it did not form at all unless the propene loading was high.

Understanding the propene/zeolite problem became a major early focus of my research program. Figure 1.2 shows the results of three complementary in situ ^{13}C NMR studies of 1-[^{13}C]propene, 2-[^{13}C]propene, and 3-[^{13}C]propene, all on zeolite

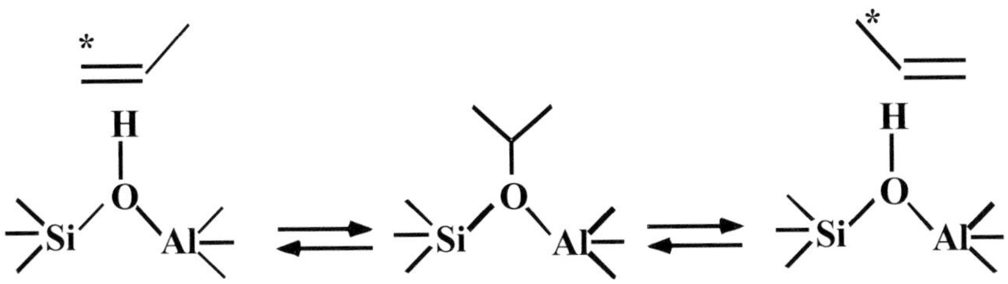

Scheme 1.1

HZSM-5. The results that we published in our 1989 follow-up paper [13] were on zeolite HY; the HZSM-5 results shown here are essentially identical, but they were obtained on a higher quality NMR instrument. The essential result in Fig. 1.2 is that as propene was heated on the zeolite it underwent $1 \rightleftarrows 3$ label exchange via an intermediate seen at $\delta = 88$ in the spectrum of 2-[^{13}C]propene, and oligomerization of propene (and other reactions) occurred after the formation of this species. The resonance at $\delta = 250$ is not seen at all in Fig. 1.2, though it readily formed with an excess of propene. Scheme 1.1 shows propene undergoing $1 \rightleftarrows 3$ label exchange by way of a framework alkoxy intermediate of the type originally predicted by Kazansky [14]. This reaction occurs without the formation of the isopropyl cation, even as a transient intermediate at a miniscule steady-state concentration. Had the isopropyl cation formed to any degree, propene would have undergone $1 \rightleftarrows 2 \rightleftarrows 3$ label exchange, because the isopropyl cation equilibrates with protonated cyclopropane.

But what of the peak at $\delta = 250$? We were able to show that it was due to the 1,2,3-trimethylcyclopentenyl carbenium ion, a very stable carbenium ion that also forms in liquid acid media through a complex pathway of oligomerization, cyclization, and hydrogen transfer. It was a spectator. Much later, we discovered that the cyclopentenyl cation is an intermediate in other reactions on zeolites [15], but its role in the first propene study was to mislead us. By the time we had it all sorted out, I was hooked on the study of reactions on solid catalysts.

References

1 STÖCKER, M., *Microporous Mesoporous Mater.* 1999, **29**, 3–48.

2 http://www.iza-structure.org/databases/

3 HWANG, S. J.; PETUCCI, C.; RAFTERY, D., *J. Am. Chem. Soc.* 1998, **120**, 4388–4397.

4 CORMA, A., *Chem. Rev.* 1995, **95**, 559–614.

5 FINK, G.; STEINMETZ, B.; ZECHLIN, J.; PRZYBYLA, C.; TESCHE, B., *Chem. Rev.* 2000, **100**, 1377–1390.

6 GRASSELLI, R. K., *Catal. Today* 1999, **49**, 141–153.

7 BARTERU, M. A., *Chem. Rev.* 1996, **96**, 1413–1430.

8 LAZO, N. D.; MURRAY, D. K.; KIEKE, M. L.; HAW, J. F., *J. Am. Chem. Soc.* 1992, **114**, 8552–8559.

9 Haw, J. F.; Nicholas, J. B.; Xu, T.; Beck, L. W.; Ferguson, D. B., *Acc. Chem. Res.* 1996, **29**, 259–267.

10 Arakawa, H.; Aresta, M.; Armor, J. N.; Barteau, M. A.; Beckman, E. J.; Bell, A. T.; Bercaw, E.; Creutz, C.; Dinjus, E.; Dixon, D. A.; Domen, K.; Dubois, D. L.; Eckert, J.; Fujita, E.; Gibson, D. H.; Goddard, W. A.; Goodman, D. W.; Keller, J.; Kubas, G. J.; Kung, H. H.; Lyons, J. E.; Manzer, L. E.; Marks, T. J.; Morokuma, K.; Nicholas, K. M.; Periana, R.; Que, L.; Rostrup-Nielson, J.; Sachtler, W. M. H.; Schmidt, L. D.; Sen, A.; Somorjai, G. A.; Stair, P. C.; Stults, B. R., Tumas, W., *Chem. Rev.* 2001, **101**, 953–996.

11 Eppler, A. S.; Zhu, J.; Anderson, E. A.; Somorjai, G. A., *Topics Catal.* 2000, **13**, 33–41.

12 Zardkoohi, M.; Haw, J. F.; Lunsford, J. H., *J. Am. Chem. Soc.* 1987, **109**, 5278–5280.

13 Haw, J. F.; Richardson, B. R.; Oshiro, I. S.; Lazo, N. L.; Speed, J. A., *J. Am. Chem. Soc.* 1989, **111**, 2052–2058.

14 Kazansky, V. B., Acc. Chem. Res. 1991, **24**, 379–382.

15 Haw, J. F.; Nicholas, J. B.; Song, W. G.; Deng, F.; Wang, Z. K.; Xu, T.; Heneghan, C. S., *J. Am. Chem. Soc.* 2000, **122**, 4763–4775.

2
In Situ Catalysis and Surface Science Methods

P. Chen and G. A. Somorjai

Abstract

Molecular level studies of surface phenomena over the past 30 years have greatly increased our knowledge of the chemical, mechanical, electrical, magnetic, and optical properties of surfaces. As a result, new technologies have been developed based on these properties of surfaces, creating wealth, new scientific knowledge, and the need for new surface instrumentation with ever increasing temporal, spatial, and energy resolution. Over 60 techniques have been developed to study surface chemistry. The most recent developments have been focused on in situ methods that can scrutinize, at the molecular level, so-called buried interfaces, as well as solid-high pressure gas, solid-liquid, and solid-solid interfaces. In this chapter, we review our recent development of two in situ surface techniques, sum frequency generation and high-pressure scanning tunneling microscopy, for studies of catalytic reactions.

2.1
Introduction

Over the past 30 years, new surface technologies have been developed that have greatly improved the quality of life by changing our lifestyle and improving our health and environment. For example, the development of catalytic converters for automobiles has greatly reduced the hazardous emissions from car exhausts. Such developments require an understanding of surface properties, structures, compositions, and dynamics, which are at the heart of surface materials based technologies. Advances in instrumentation for studying surfaces with nanoscale spatial resolution and continuous improvements in temporal and spectral resolution provide opportunities to develop new technology. In turn, the ever-present need for improved technology provides opportunities for research to advance the frontiers of molecular surface science.

In this chapter, we review the state of the art of in situ surface science techniques and discuss the frontier areas of research in catalysis. First, the tools of surface studies are reviewed with special emphasis on those techniques that can be employed to monitor the working surface in gases at atmospheric pressure, in liquids, or when in contact with other solid surfaces (so-called buried surfaces). Thereafter, we discuss the applications of these novel in situ techniques in studies of catalytic reactions. Finally, we point to the new directions of surface materials research that will likely provide the foundation for discovering new surface technologies.

2.2
Surface Science Tools

The development of molecular surface science depends on surface characterization techniques. Over the last three decades, numerous techniques have been developed to study various surface properties, including structure, composition, oxidation states, and changes in chemical, electronic, and mechanical properties based on the scattering, absorption, or emission of photons, electrons, atoms, or ions. Most surface probes require low pressures ($<10^{-3}$ Torr) or UHV (10^{-9} Torr) for their application because of the large cross-sections for scattering that prevent the particles on surface probes from reaching the detectors at higher pressures. To overcome this limitation, high-pressure reaction cells were constructed and placed inside ultra high vacuum systems equipped with different surface probes [1]. The sample to be analyzed was first prepared and characterized in the UHV chamber, and then transferred to the high-pressure cell, where it was subjected to the usual high-pressure and/or high-temperature conditions. Afterwards, the sample was

Tab. 2.1 Molecular surface characterization techniques operating under high-pressure reaction conditions

Isotope exchange
In situ NMR
Fourier-transform infrared spectroscopy
In situ Raman spectroscopy
Sum frequency generation surface vibrational spectroscopy
Positron emission ^{11}C, ^{13}N, and ^{15}O
In situ Mössbauer spectroscopy
In situ scanning tunneling microscopy
In situ electron microscopy
In situ X-ray absorption spectroscopy

transferred, without exposure to air, to the evacuated UHV chamber, where the surface probes were located for subsequent surface analysis.

During the past ten years, several surface techniques (Tab. 2.1) have become available that may be employed at high ambient pressures to monitor the behavior of small surface area (~1 cm^2) samples, providing in situ atomic and molecular level information with high surface sensitivity. Among these techniques, sum frequency generation (SFG)-surface vibrational spectroscopy and scanning tunneling microscopy (STM) are the two most often utilized techniques in our laboratory. Both of these techniques can be performed over a 14 order of magnitude pressure range (10^{-10}–10^4 Torr) without a significant change in signal quality in terms of spatial or energy resolution. Using these two techniques, both substrate and adsorbate structures can be monitored during reactions carried out at high pressures. Since these two techniques are relatively new to catalysis, we will describe them in some detail.

2.2.1
Sum Frequency Generation (SFG)-surface specific vibrational spectroscopy

With the advent of high-energy, pulsed picosecond lasers, nonlinear optical techniques such as sum frequency generation have been shown to be very powerful methods for in situ investigations of surfaces and interfaces under catalytic conditions. In an SFG experiment, two input beams are temporally and spatially overlapped on a surface to generate a third beam at the sum of the two input frequencies. Normally, one input beam is a visible green beam with a fixed wavelength of 532 nm, while the other is tunable in the infrared region allowing probing of the vibrational modes of surface species. The process is shown schematically in Fig. 2.1.

During the SFG process, an IR photon induces a vibrational transition from a ground state to an excited state. Simultaneously, a green photon induces a transition to a higher energy virtual state through a Raman process. This high-energy virtual state relaxes by emitting a photon with a frequency equal to the sum of the two input photon frequencies, and the angle of emission is determined by the phase-matching condition. The output beam is therefore in the visible region and is easily detected. The intensity of the sum frequency output reaches a maximum when the IR frequency is near a vibrational resonance, and so a vibrational spectrum can be acquired by tuning the IR beam over the frequency range of interest. In order for a vibrational mode to be SFG-active, both IR and Raman selection rules must be satisfied. From the electric dipole approximation, SFG is not allowed in centrosymmetric media such as bulk materials and gases. Therefore, SFG is a surface-specific technique since the process can only occur at surfaces or interfaces where the centrosymmetry is broken [2–6].

Sum Frequency Generation (SFG)

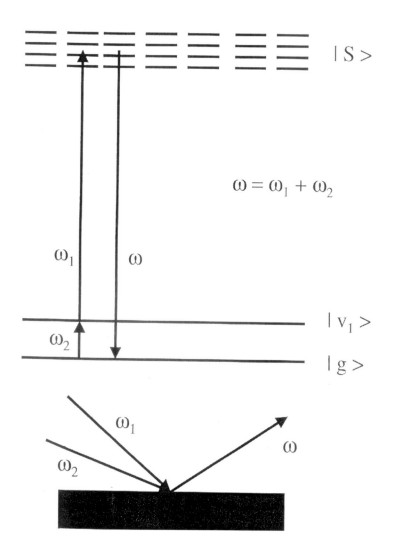

Fig. 2.1 Schematic representation of the sum frequency generation (SFG) technique

2.2.2
The high-pressure high-temperature Scanning Tunneling Microscope (STM)

Scanning tunneling microscopy (STM) is a valuable technique for experiments in high-pressure environments simply because it employs a probe that is positioned just a few Å from the surface. The proximity of the probe and sample allows us to ignore the gas-phase environment so that we can concentrate on the first few metal surface layers and the adsorbate layer.

We have developed an apparatus that combines a high-pressure, high-temperature STM with a UHV surface analysis chamber as shown in Fig. 2.2. The STM portion of the apparatus contains the STM stage and the STM manipulator. A UHV chamber and load lock system are attached to the STM chamber with gate valves and the whole apparatus is floated on "air legs" for vibration isolation.

The STM scan head is a Beetle-type STM equipped with sample heating/cooling and in situ probe tip exchange. The STM chamber can be isolated from the rest of the system by gate valves, allowing the introduction of high gas pressures, while the UHV chamber and the load lock remain under vacuum. The STM tip can remain in tunneling range as the pressure in the chamber is increased, so that the same area of the surface can be imaged before and after the gas is introduced.

In order to heat the sample at high pressures (greater than 10^{-4} Torr), a small, 360 W tungsten filament halogen-filled quartz bulb is used. This provides radia-

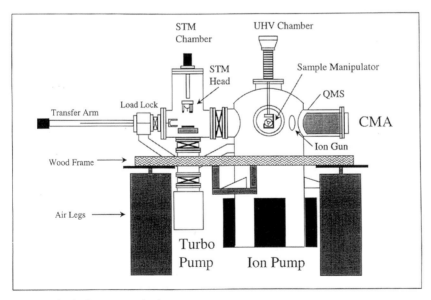

Fig. 2.2 The high-pressure, high-temperature scanning tunneling microscope. Pressure range: 5×10^{-10} Torr to 1 atm. Temperature range: 25 to 400°C

tive heating without making mechanical contact with the sample, preserving the vibrational isolation of the STM stage. This heating system can achieve temperatures of greater than 900 K at all pressures up to 750 Torr.

The chemical and thermal stability of the scanning probe tip is of primary importance under the high-pressure and high-temperature conditions of our studies. Choosing an acceptable tip material is the most challenging aspect of high-pressure STM, which must be resolved for the application of STM to chemical reaction studies. For example, tungsten tips allow stable imaging at room temperature in high-pressure CO, but they are not stable in oxygen or mixtures of CO and oxygen. Probe tips of platinum and platinum alloys (Pt/Rh, Pt/Ir, Pt/Au) are stable in high-pressure oxygen and hydrogen, but are unstable in CO. Electrochemically sharpened gold tips provide the chemical stability required for STM studies in high-pressure gas mixtures, but they are susceptible to blunting, especially at elevated temperatures.

2.3
Applications of In Situ Methods in Surface Science to Catalysis

2.3.1
High-Pressure SFG Studies

2.3.1.1   Ethylene hydrogenation on Pt(111)

The first mechanistic proposals concerning the hydrogenation of the simplest olefin, ethylene, on a platinum surface were made by Horiuti and Polanyi in the 1930's [7]. Their model proposed that ethylene is adsorbed onto a clean platinum surface by breaking one of the double bonds and then forming two σ-bonds with the metal surface. This intermediate is known as di-σ-bonded ethylene. One of the metal-carbon bonds is then envisaged as being hydrogenated with adsorbed hydrogen creating an ethyl intermediate, allowing ethane production by final hydrogenation of the remaining metal-carbon bond. Surface techniques such as ultraviolet photoemission spectroscopy (UPS) were used under UHV conditions to study the mechanism of ethylene hydrogenation on platinum single crystals. It was shown that below 52 K, ethylene physisorbs through the π-bond, this being referred to as π-bonded ethylene [8]. As the temperature is increased beyond 52 K, the π-bond is broken and di-σ-bonded ethylene is formed [9]. Ethylidyne (M≡CCH$_3$) is formed as di-σ-bonded ethylene dehydrogenates and transfers a hydrogen atom from one carbon atom to the other [10]. As the surface is heated further, ethylidyne decomposes into graphitic precursors [11].

Ethylidyne is not believed to be a reaction intermediate in ethylene hydrogenation. Davis and coworkers [12] showed ethylidyne hydrogenation to be several orders of magnitude slower than the overall hydrogenation of ethylene to ethane. Furthermore, Beebe and coworkers, using in situ infrared transmission spectroscopy, showed that the reaction rate was the same on a Pd/Al_2O_3-supported catalyst, irrespective of whether or not the surface was covered with ethylidyne [13]. These studies indicate that ethylidyne is a spectator species and is not directly involved in ethylene hydrogenation.

To determine the importance of π-bonded ethylene and di-π-bonded ethylene, Mohsin and coworkers, using transmission infrared spectroscopy, showed that both species are hydrogenated on a supported Pt/Al_2O_3 catalyst as hydrogen is passed over the surface [14]. Moreover, Mohsin and coworkers also showed that only di-σ-bonded ethylene is converted to ethylidyne when the catalyst is heated in the absence of hydrogen. These studies, however, were not performed under actual high-pressure reaction conditions because gas-phase ethylene interferes with infrared experiments. Since gas-phase ethylene does not generate an SFG signal, SFG is an ideal technique for studying ethylene hydrogenation in situ under high-pressure reaction conditions.

High-pressure ethylene hydrogenation experiments have been monitored in situ by SFG under hydrogen pressures between 2 and 700 Torr [15]. In each experiment, the ethylene pressure was kept below the hydrogen pressure, and the ethylene pressures were kept in the regime where the reaction order would be zero with respect to ethylene. The results shown here are for experiments in which 35 Torr of ethylene, 100 Torr of hydrogen, and 625 Torr of He were introduced at 295 K. Figure 2.3a shows the SFG spectrum, from which the presence of several surface species is apparent. Ethylidyne is observed at 2878 cm^{-1}, di-σ-bonded ethylene is observed at 2910 cm^{-1}, and πbonded ethylene gives rise to two broad features around 3000 cm^{-1}. The intensity of the peaks corresponds to 0.15 ML, 0.08 ML, and 0.04 ML for ethylidyne, di-σ-bonded ethylene, and π-bonded ethylene, respectively. The spectrum shown in Fig. 2.3a remained unchanged for hours, indicating that the composition of adsorbates remained essentially the same on the surface over the lifetime of the experiment. As SFG spectra were recorded, the gas-phase composition was simultaneously monitored by gas chromatography. Analysis of the GC data showed a turnover rate (TOR) of 11 ± 1 ethylene molecules converted per surface platinum atom per second.

After the batch reactor was evacuated, another SFG spectrum was recorded (Fig. 2.3b), which showed that only ethylidyne remained on the surface. The peak intensity also indicated that the amount of ethylidyne absorbed onto the surface increased to saturation coverage (0.25 ML). This change in surface coverage was not unexpected, as the UHV experiments showed that di-σ-bonded ethylene hydrogenates to ethylidyne at 300 K when hydrogen is not present. This result confirms that di-σ-bonded ethylene and π-bonded ethylene can only exist at 300 K when hydrogen is present.

(a)

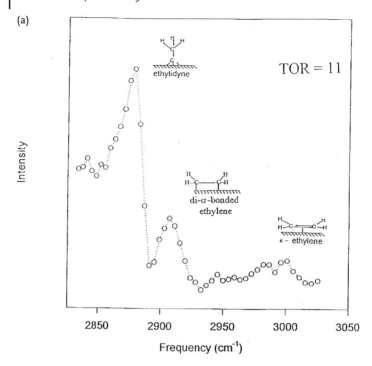

(b)

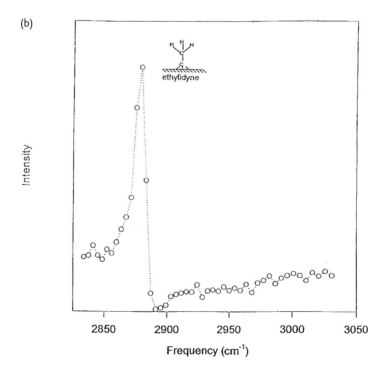

(c)

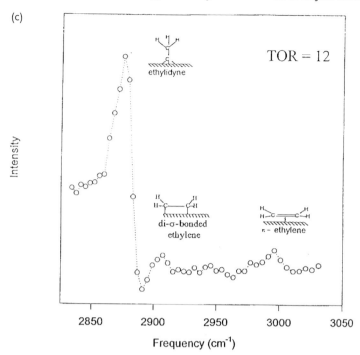

ethylidyne

TOR = 12

di-σ-bonded
ethylene

π - ethylene

Intensity

2850 2900 2950 3000 3050

Frequency (cm⁻¹)

Fig. 2.3 (a) SFG spectrum of ethylene hydrogenation at 295 K on Pt(111); 100 Torr H₂, 35 Torr ethylene, and 615 Torr He. Species observed under these reaction conditions include ethylidyne di-σ-bonded ethylene, and π-bonded ethylene. (b) SFG spectrum after the reaction cell was evacuated. Only strongly bound ethylidyne is observed. (c) SFG spectrum under the same conditions as for (a) on Pt(111) preadsorbed with 0.25 ML of ethylidyne.

The spectrum shown in Fig. 2.3c was acquired during a similar high-pressure experiment, but in this case the surface was pretreated in UHV with 4 L of ethylene at 295 K. This experiment was performed to determine the effect of a saturation coverage of ethylidyne during a reaction. After the pretreatment with ethylene, 35 Torr of ethylene, 100 Torr of H₂, and 625 Torr of He were introduced into the batch reactor. The vibrational spectrum is considerably different to that shown in Fig. 2.3a. Thus, ethylidyne appears at a higher coverage than in the previous experiment, while the di-σ-bonded feature is much smaller (0.02 ML compared to 0.08 ML). On the other hand, the π-bonded ethylene appears at a similar coverage as seen in Fig. 2.3a. Gas chromatographic data revealed a turnover rate of 12 ± 1 molecules per platinum site per second, which is almost identical to the turnover rate in the previous experiment.

From the fact that the rate of the reaction remained the same while the amount of di-σ-bonded ethylene on the surface decreased, it becomes apparent that di-σ-bonded ethylene is not an important reaction intermediate in ethylene hydrogenation. Furthermore, it appears that ethylidyne and di-σ-bonded ethylene compete for sites. Once ethylidyne forms on the surface, the subsequent adsorbtion of di-σ-bonded ethylene is blocked. From previous work, it is known that ethylidyne is not a reaction intermediate. This study indicates that di-σ-bonded ethylene is not a reaction intermediate either.

Because the concentration of π-bonded ethylene was the same in both cases and because the reaction rate remained the same, it appears that π-bonded ethylene is probably the key reaction intermediate in ethylene hydrogenation on Pt(111).

Additional high-pressure ethylene hydrogenation reactions were performed, in which the hydrogen and ethylene pressures were increased compared to those in the previous case. A clean Pt(111) crystal was exposed to 723 Torr of hydrogen and 60 Torr of ethylene at 295 K. In this case (Fig. 2.4), the SFG spectrum showed that the surface was not saturated with ethylidyne as in the previous experiment at 100 Torr of hydrogen. Little change was seen in the size of the π-bonded ethylene peak, but the peaks associated with di-σ-bonded ethylene were decreased compared to Figs. 2.3a and 2.3c. Two new peaks at 2850 and 2925 cm^{-1} were also observed, similar to the peaks observed in the UHV experiment (not shown), and

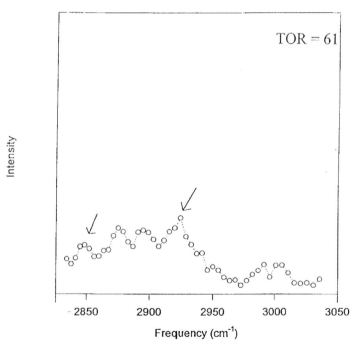

Fig. 2.4 SFG spectrum of ethylene hydrogenation on Pt(111) at 295 K; 727 Torr H$_2$ and 60 Torr of ethylene. Ethyl species are observed, as indicated by the two arrows

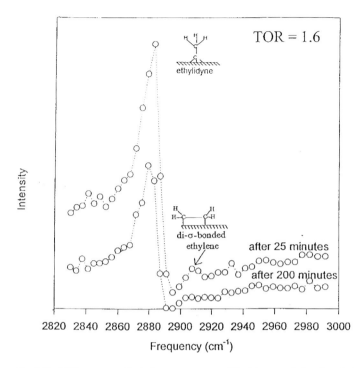

Fig. 2.5 SFG spectra of ethylene hydrogenation on Pt(111) at 295 K after 25 minutes and 200 minutes; 1.75 Torr H$_2$, 0.25 Torr ethylene, and 758 Torr argon.

Ethylidyne and di-σ bonded ethylene are observed after 25 minutes, but only ethylidyne is observed after 200 minutes.

thus correspond to an ethyl species. Gas chromatographic data revealed a turnover rate of 61 ± 3 molecules per platinum site per second.

At lower hydrogen and ethylene pressures (1.75 Torr of H$_2$, 0.25 Torr of ethylene, and 758 Torr of Ar), the spectrum (Fig. 2.5) showed that the surface was saturated with ethylidyne and only a small amount of di-σ-bonded ethylene was observed. As the reaction progressed, the di-σ-bonded ethylene peak decreased until it was no longer observable after 200 minutes. At this time, ethylidyne was the only observable species on the surface and at no time during the reaction was π-bonded ethylene observed. The turnover rate throughout the entire reaction was 1.6 molecules per platinum site per second.

From these high-pressure ethylene hydrogenation experiments using both (111) and (100) platinum crystal surfaces, in which the gas-phase composition was monitored by GC at the same time as SFG spectra were being acquired, it was observed that the ethylene hydrogenation turnover rate was independent of the surface concentration of di-σ-bonded ethylene or the ethylidyne to di-σ-bonded ethylene ratio. It is thus suggested that ethylidyne and di-σ-bonded ethylene are merely spectator

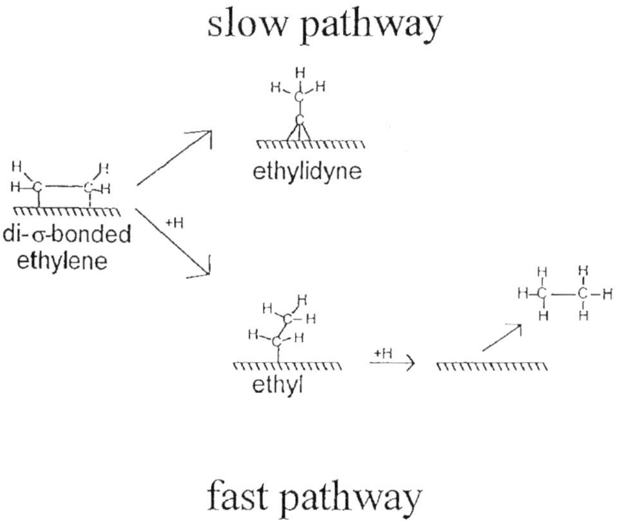

slow pathway

Fig. 2.6 Proposed reaction pathway for ethylene hydrogenation. The slow pathway proceeds through an ethylidyne species and the fast pathway proceeds through an ethyl intermediate.

species in the reaction and that the most likely reaction intermediate is π-bonded ethylene. The surface coverage of π-bonded ethylene was determined to be about 4% of a monolayer, suggesting that the turnover rate is actually 25 times higher if quoted per surface intermediate than if quoted per platinum atom. These results indicate that not all platinum atoms on the surface are active at any given time. Studies have shown that ethylene hydrogenation is a structure-insensitive reaction and, therefore, the specific location at which key reaction steps take place must be at sites that are available on all crystallographic planes of platinum [16].

From the data presented above, the following reaction pathway for ethylene hydrogenation is proposed. Hydrogen dissociatively chemisorbs on a clean or ethylidyne-covered platinum surface. Ethylene then physisorbs to form π-bonded ethylene and is subsequently hydrogenated in a stepwise manner to form an ethyl intermediate en route to ethane production. The reaction pathway is outlined in Fig. 2.6.

2.3.1.2 Propylene hydrogenation and dehydrogenation on Pt(111)

It has been shown that propylene adsorbs on the Pt(111) surface as a di-σ-bonded species in UHV below 220 K [17]. As with ethylene, once the surface is heated, the di-σ-bonded propylene species converts to propylidyne at around 300 K with the evolution of hydrogen [18]. Above 300 K, propylidyne decomposes into graphitic precursors.

Kinetic data show that propylene hydrogenation at atmospheric pressure on platinum is 0.5 order in hydrogen and zero order in propylene at 300 K [19,20]. Examination of supported Pt/SiO_2 after exposure to propylene and hydrogen reveals the presence of several species: propylidyne, di-σ-bonded propylene, and π-bonded propylene [21]. There are two proposed pathways for the stepwise hydrogenation of propylene. The reaction may proceed through either a 2-propyl intermediate or a 1-propyl intermediate depending on whether the first hydrogen is added to the terminal carbon or the internal carbon of the olefin.

High-pressure catalytic reactions were performed while acquiring SFG spectra and simultaneously monitoring the gas-phase composition in the batch reactor by gas chromatography. Under the highest-pressure conditions, 723 Torr of hydrogen and 40 Torr of propylene were introduced into the batch reactor at 295 K. The SFG spectrum (Fig. 2.7a) featured seven different peaks. The largest peak seen at 2830 cm^{-1} can be assigned to a 2-propyl group. The peaks at 3050 cm^{-1} and 2960 cm^{-1} correspond to π-bonded propylene and propylidyne, respectively. The peaks between 2860 and 2940 cm^{-1} are more difficult to assign, but may be attributed to 1-propyl, 2-propyl, and/or di-σ-bonded propylene. Under these reaction

(a)

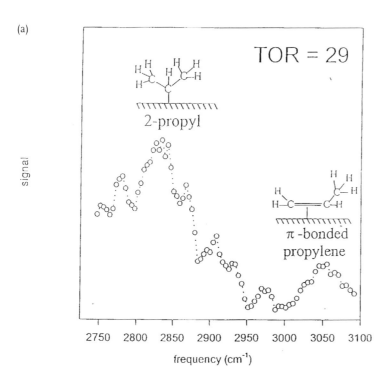

signal

frequency (cm^{-1})

(b)

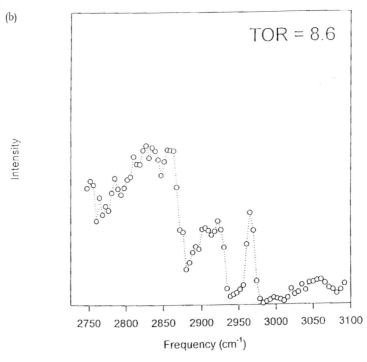

(c)

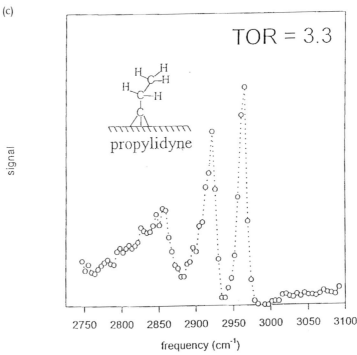

(d)

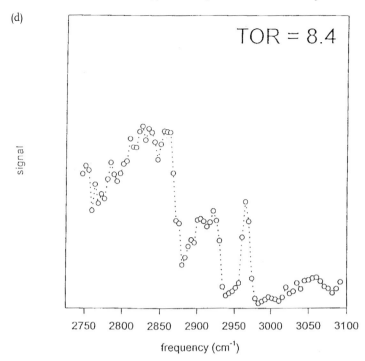

Fig. 2.7 (a) SFG spectrum of propylene hydrogenation on Pt(111) at 295 K with 723 Torr H_2 and 40 Torr propylene. (b) SFG spectrum of propylene hydrogenation on Pt(111) at 295 K with 100 Torr H_2, 40 Torr propylene, and 617 Torr He. (c) SFG spectrum of propylene hydrogenation on Pt(111) at 295 K with 13.5 Torr H_2, 5 Torr propylene, and 746 Torr He. (d) SFG spectrum of propylene hydrogenation on Pt(111) preadsorbed with propylidyne at 295 K with 100 Torr H_2, 40 Torr propylene, and 617 Torr He

conditions, the turnover rate was 29 propane molecules formed per exposed platinum atom per second. Surprisingly, in spite of the fact that the intensity of the peak due to the di-σ-bonded propylene species should be lower than that of the peak due to the σ-bonded propylene species because of the metal dipole selection rules and the SFG selection rules, the π-bonded propylene peak is actually larger than that of the di-σ-bonded propylene species. This evidence suggests that the concentration of the π-bonded species is higher than that of the di-σ-bonded propylene species.

Another interesting point about the spectrum is the fact that the intensity of the peak associated with the 2-propyl species is at least three times greater than that of the peak for the 1-propyl species. This also indicates that the 2-propyl species was present at a higher concentration than the 1-propyl species, as the UHV experiments showed the 1-propyl species to have a higher SFG cross-section.

When the reaction was performed at 100 Torr of hydrogen, 40 Torr of propylene, and 617 Torr of He at 295 K, both the SFG spectrum (Fig. 2.7b) and the turnover rate were seen to be dramatically different from in the previous high-pressure case. The turnover rate decreased to 8.6 molecules per platinum site per second, and peaks were seen at 2960 cm^{-1} and 2920 cm^{-1}, which can be assigned to propylidyne. Moreover, the SFG features at 3050 cm^{-1} and 2830 cm^{-1} associated with π-bonded propylene and 2-propyl, respectively, were considerably weaker. However, the feature at 2863 cm^{-1} was more intense, indicating a possible net reorientation of the 2-propyl species. The feature just below 2900 cm-1 also intensified and is assigned to di-σ-bonded propylene. The increase in the propylidyne and di-σ-bonded propylene peaks, along with the decrease in turnover rate, allows us to rule out these species as important reaction intermediates in propylene hydrogenation.

When the pressures of the reactants were further decreased (13.5 Torr of hydrogen, 5 Torr of propylene, 746 Torr He), the turnover rate decreased to 3.3 molecules per site per second. This rate is nearly an order of magnitude lower than in the high-pressure case discussed above, and the corresponding SFG spectrum was dominated by large features associated with propylidyne at 2850 cm^{-1}, 2915 cm^{-1}, and 2960 cm^{-1}. The broad band near 2850 cm^{-1} was probably due to a small concentration of 2-propyl.

To determine the effect of propylidyne on the surface under reaction conditions, a clean Pt(111) sample was exposed to 4 L of propylene at 295 K forming a saturation coverage of propylidyne. The sample was then subjected to 40 Torr of propylene, 100 Torr or hydrogen, and 760 Torr of He. The turnover rate was found to be 8.4 molecules per site per second, which is nearly identical to the turnover rate associated with Fig. 2.7b. The SFG spectrum (Fig. 2.7c) was also very similar to those observed in UHV (not shown). This demonstrates that the amount of propylidyne adsorbed on the surface quickly attains equilibrium during the hydrogenation process regardless of how much is initially adsorbed onto the surface.

From the SFG experiments outlined above, it is clear that there are two important reaction intermediates in propylene hydrogenation. Analogous to ethylene hydrogenation, propylene adsorbs as a π-bonded species onto the platinum surface. This π-bonded species can then hydrogenate to form either 1-propyl or 2-pro-

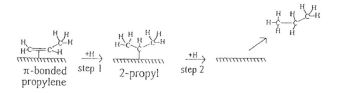

Fig. 2.8 Proposed reaction pathway for propylene hydrogenation on Pt(111). The important reaction intermediates are π-bonded propylene and 2-propyl

pyl moieties. The SFG spectra indicate that the 2-propyl species is present at a higher concentration than the 1-propyl species. This is reasonable, as the bond between an internal carbon and the underlying metal should be slightly weaker than a carbon-metal bond involving a terminal carbon. With this data in mind, it is believed that propylene hydrogenation proceeds more rapidly through a 2-propyl intermediate than through a 1-propyl intermediate. The resulting proposed reaction pathway is outlined in Fig. 2.8.

2.3.1.3 Cyclohexene hydrogenation and dehydrogenation on Pt(111) and Pt(100)

Several spectroscopic techniques have been utilized to study cyclohexene adsorbed on Pt(111) and Pt(100) under UHV conditions. These techniques include HREELS, EELS, and IRAS [22–26]. The results of these studies show that cyclohexene dehydrogenates to a C_6H_9 intermediate en route to benzene formation on Pt surfaces. However, these techniques are limited to low-pressure studies. To study the actual reaction pathway under catalytic reaction conditions, high-pressure experiments are required to provide much needed structural information.

Possible important intermediates in cyclohexene high-pressure catalytic reactions may be 1,3-cyclohexadiene and 1,4-cyclohexadiene, hence it is important to obtain SFG spectra of these species on platinum under high-pressure conditions. High-pressure catalytic hydrogenation experiments of both 1,3-cyclohexadiene and 1,4-cyclohexadiene were carried out on a Pt(111) sample at 295 K. The turnover rate for 1,3-cyclohexadiene was very fast at 100 molecules per site per second, while the hydrogenation of 1,4-cyclohexadiene was much slower at only five molecules per site per second. The SFG spectra for these reactions are shown in Figs. 2.9 and 2.10. These data are important for determining the key reaction intermediates during cyclohexene hydrogenation and dehydrogenation on both Pt(111) and Pt(100).

For high-pressure cyclohexane hydrogenation and dehydrogenation reactions, 10 Torr of cyclohexene, 100 Torr of hydrogen, and 650 Torr of He were introduced into the high-pressure cell [27]. The sample was heated at between 300 K and 600 K while spectroscopic changes of surface species were monitored by SFG and the gas composition was monitored by gas chromatography (GC). Turnover rates were calculated on the basis of the GC data at each temperature and are shown in Fig. 2.11a and Fig. 2.11b for cyclohexene on Pt(111) and Pt(100), respectively. At 300 K, the hydrogenation and dehydrogenation turnover rates are negligible. Upon heating the crystal sample, the hydrogenation turnover rate increases and reaches a maximum of 78 molecules per site per second at 400 K for Pt(111), and a maximum of 38 molecules per site per second at 425 K for Pt(100). Near the temperature of the maximum hydrogenation turnover rate on both crystals, the dehydrogenation turnover rate begins to increase with temperature while the hydrogenation rate begins to decrease with increasing temperature. On the Pt(111) surface, the maxi-

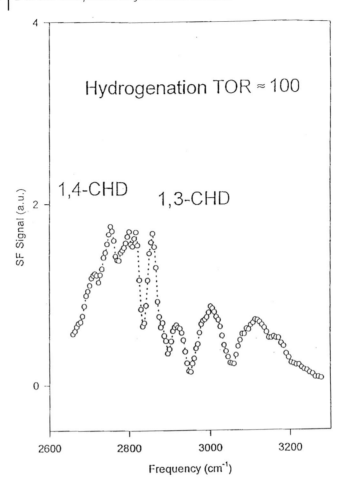

Fig. 2.9 SFG spectrum of 1,4-cyclohexadiene hydrogenation on Pt(111). Both 1,4- and 1,3-cyclohexadiene are observed

mum dehydrogenation rate is 58 molecules per site per second at 475 K, while on the Pt(100) surface the maximum dehydrogenation turnover rate is 75 molecules per site per second at 500 K.

Assuming that the dehydrogenation turnover rate is negligible in the region of increasing hydrogenation turnover rate, the activation energy for cyclohexene hydrogenation is calculated to be 8.9 and 15.5 kcal mol^{-1} for Pt(111) and Pt(100), respectively. As the temperature is increased above that of the cyclohexene hydrogenation rate maximum, the hydrogenation rate decreases as the dehydrogenation rate increases. This may indicate that the numbers of active surface sites for the

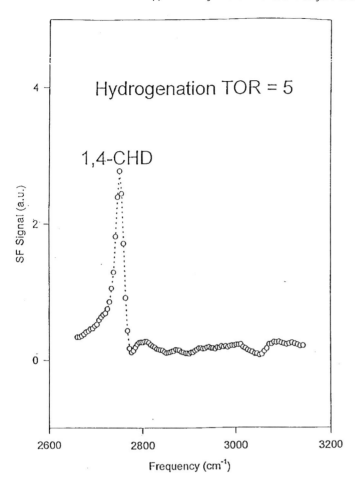

Fig. 2.10 SFG spectrum of 1,4-cyclohexadiene
hydrogenation on Pt(111). Only 1,4-cyclohexadiene is observed

two processes change. Otherwise, the hydrogenation reaction would obey the Arrhenius law over the entire temperature range.

While the catalytic reactions were being performed, SFG spectra were collected simultaneously. Figure 2.12 shows the SFG spectra obtained at three different temperatures: that at which the gases were introduced into the chamber, that of the maximum hydrogenation turnover rate, and that of the maximum dehydrogenation rate. On Pt(111) (Fig. 2.12a), a sharp feature at 2755 cm^{-1} is observed at 295 K. The SFG spectrum (Fig. 2.10) from the high-pressure reaction of 1,4-cyclohexadiene allows the assignment of the 2755 cm^{-1} feature to 1,4-cyclohexadiene. From the high-pressure hydrogenation of 1,3-cyclohexadiene (Fig. 2.9), the weaker

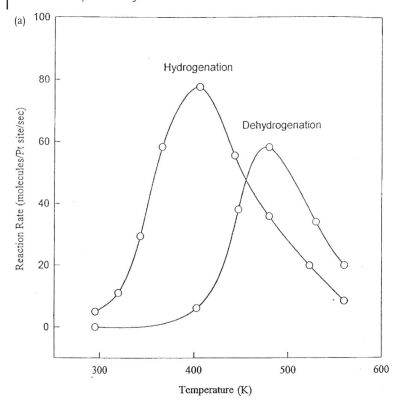

(a)

features at higher frequencies in Fig. 2.12a can be attributed to 1,3-cyclohexadiene. It is important to note that there is no evidence of a c-C_6H_9 intermediate species as observed under UHV conditions. On Pt(100) (Fig. 2.12b) at 300 K, the major prominent features can be assigned to 1,3-cyclohexadiene, while a weak feature at around 2780 cm^{-1} can be assigned to 1,4-cyclohexadiene.

At the maximum hydrogenation rate, which occurs at 403 K and 425 K for Pt(111) and Pt(100), respectively, the two similar SFG spectra indicate the presence of 1,3-cyclohexadiene as the major species on the surface. This spectral evidence indicates that cyclohexene hydrogenation proceeds through a 1,3-cyclohexadiene intermediate. As the temperature is increased to that of the maximum dehydrogenation rate, the SFG spectra for the two crystal surfaces become different once more. On Pt(111), the spectrum indicates the presence of both 1,3- and 1,4-cyclohexadiene, whereas on Pt(100) the major spectral features are assigned only to 1,3-cyclohexadiene. Considering the differences in the two SFG spectra and the difference in the turnover rates for maximum dehydrogenation, a reaction pathway is proposed for dehydrogenation on both Pt surfaces. On Pt(111), cyclohexene dehydrogenation can proceed through either a 1,3- or 1,4-cyclohexadiene inter-

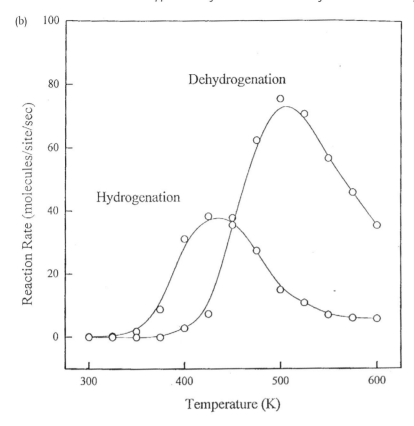

Fig. 2.11 (a) The temperature dependence of the turnover rate of hydrogenation and dehydrogenation reactions of cyclohexene on Pt(111); 10 Torr cyclohexene, 100 Torr H_2, and 650 Torr He. (b) The temperature dependence of the turnover rate of hydrogenation and dehydrogenation reactions of cyclohexene on Pt(100)

mediate, whereas on Pt(100) dehydrogenation only occurs through a 1,3-cyclohexadiene intermediate. Fig. 2.13 shows the different reaction pathways for the two surfaces.

Because cyclohexene dehydrogenation occurs more rapidly on Pt(100) than on Pt(111), and because 1,4-cyclohexadiene is absent from the Pt(100) surface, it is believed that 1,4-cyclohexadiene inhibits dehydrogenation on the Pt(111) surface. This is reasonable considering that 1,4-cyclohexadiene must presumably first isomerize to 1,3-cyclohexadiene before it completely dehydrogenates to form benzene.

(a)

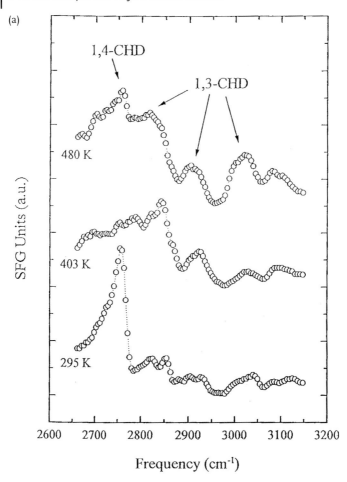

2.3.1.4 **CO oxidation on Pt(111)**

One of the most studied surface-catalyzed reactions is the oxidation of carbon monoxide to carbon dioxide over transition metals. UHV studies reveal that the thermal excitation of adsorbed oxygen initiates the reaction and IR spectroscopic studies show that atop CO is the major carbon monoxide species present on the surface [28]. It has been shown that the reaction proceeds by a Langmuir-Hinshelwood mechanism (reaction following adsorption of reactants) [29]. Under UHV, studies have shown that in order for the reaction to be initiated, the adsorbed oxygen must first be thermally excited [30]. Also, the desorption of CO_2 is dependent on the adsorption state of the oxygen species. The mobile oxygen atoms on the surface that approach and react with CO molecules are possibly responsible for CO_2 production. Infrared studies under UHV have shown that atop CO is the major CO species present on the surface under oxidative conditions.

(b)

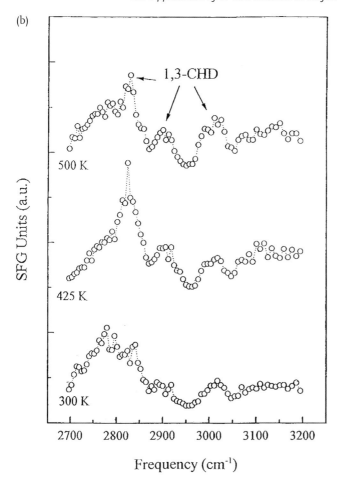

Fig. 2.12 (a) The temperature dependence of SFG spectra during high-pressure cyclohexene hydrogenation and dehydrogenation on Pt(111); 10 Torr cyclohexene, 100 Torr H$_2$, and 650 Torr He. At the maximum dehydrogenation temperature, both 1,3- and 1,4-cyclohexadiene species are observed. (b) The temperature dependence of SFG spectra during high-pressure cyclohexene hydrogenation and dehydrogenation on Pt(100). Only 1,3-cyclohexadiene is observed

The majority of what is known about CO oxidation has come from UHV studies, but the study of CO oxidation under high pressures of CO and O$_2$ using high-pressure techniques such as SFG is of practical importance. To study the high-pressure CO oxidation reaction on a Pt(111) surface, 100 Torr of CO, 40 Torr of O$_2$, and 600 Torr of He were introduced into the batch reactor [31]. Figure 2.14 shows the

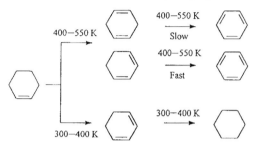

Reaction Pathway on Pt(100)

Reaction Pathway on Pt(111)

Fig. 2.13 Proposed reaction pathways for cyclohexene hydrogenation and dehydrogenation on Pt(111) and Pt(100). The absence of 1,4-cyclohexadiene during dehydrogenation on Pt(100) explains the structure sensitivity of the reaction

SFG spectra obtained at various temperatures. Below 740 K, only atop bonded CO at 2090 cm^{-1} and a low-frequency shoulder representing the incommensurate CO overlayer formation are observed [32]. As the temperature is increased to 1100 K, the intensity of the peak due to atop CO species decreases, which indicates that the concentration of atop CO decreases. Simultaneously, the intensity of the peak due to the incommensurate CO overlayer increases. The ignition temperature under these conditions is 760 K, at which the reaction becomes self-sustained, proceeding at a high constant temperature due to its high exothermicity. Comparing the SFG spectra above and below the ignition point, it is clear that they are completely different. A low-frequency broad band and peak at 2045 cm^{-1}, which dominate the spectrum at 1100 K, are assigned to incommensurate CO and CO adsorbed to defect sites. The reaction rate is found to increase with an increase in the intensity of this band.

Another high-pressure CO oxidation experiment was performed, but with O_2 rather than CO being present in excess. In this experiment, 100 Torr of O_2, 40 Torr of CO, and 600 Torr of He were introduced into the chamber. The SFG spectra obtained at various temperatures under these conditions are shown in Fig. 2.15. At low temperature, the reaction rate is low and the SFG spectrum is dominated by the atop CO species, similar to what is seen when CO is present in excess. At 540 K, a turnover rate of 28 molecules of CO_2 per platinum surface site per second is obtained. With increasing temperature, the atop CO peak intensity decreases, but the

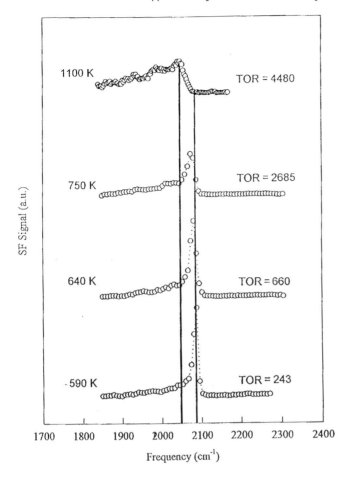

Fig. 2.14 SFG spectra of high-pressure CO oxidation on Pt(111) at temperatures above and below the ignition temperature; 40 Torr O_2, 100 Torr CO, and 600 Torr He

reaction rate increases. The ignition temperature is near 600 K, above which the system reaches the high reactivity regime and the reaction becomes self-sustained. As in the excess CO experiment, the SFG spectra above and below the ignition point are considerably different. The atop CO peak that is the dominant feature below the ignition temperature is not observed at all above the ignition point. The only new feature of note is again at 2045 cm^{-1}, as seen in the excess CO case. Two other peaks centered at 2130 cm^{-1} and 2240 cm^{-1} are also observed, but on inspection these are seen not to be real. In the high temperature regime, the nonlinear background is higher than in the low temperature regime, and the observed peaks at 2130 cm^{-1} and 2240 cm^{-1} are actually caused by a decrease in the intensity of the

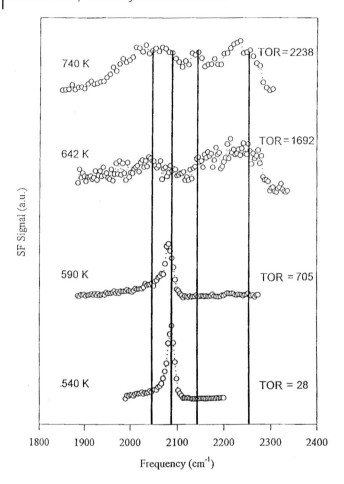

Fig. 2.15 SFG spectra of high-pressure CO oxidation on Pt(111) at temperatures above and below the ignition temperature; 100 Torr O_2, 40 Torr CO, and 600 Torr He

IR beam. The decrease in the IR power is caused by gas-phase CO in the chamber and atmospheric CO_2 outside the chamber.

Figs. 2.16a–c show the correlation between the reaction rate and the relative CO coverage of various surface species. As Fig. 2.16a indicates, when the reaction rate increases, the intensity of the peak due to atop CO species decreases. This clearly indicates that atop CO is not the active participant in CO oxidation. Its presence on the surface may actually inhibit the reaction. Figs. 2.16b and 2.16c show that the reaction rate is proportional to the concentration of the incommensurate CO species both above and below the ignition temperature. These data show that this incommensurate surface species must be directly responsible for CO oxidation at all temperatures.

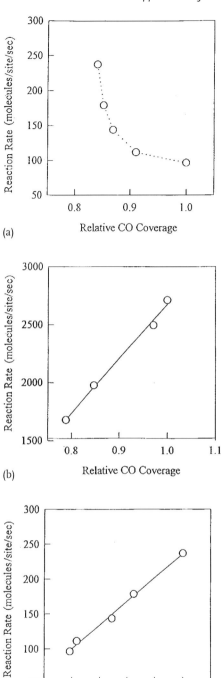

(a)

(b)

(c)

Fig. 2.16 (a) Reaction rate versus atop CO coverage at 590 K. This evidence shows that atop CO is not the active species in CO oxidation. (b) Reaction rate versus surface coverage of incommensurate CO at 720 K. (c) Reaction rate versus surface coverage of incommensurate (CO) at 590 K

2.3.2
High-Pressure STM Studies

2.3.2.1 High-pressure CO on Pt(111) [33]

As a first high-pressure STM experiment, we studied high-pressure CO on a Pt(111) surface. In a typical experiment, the sample was cleaned, transferred to the STM chamber, and imaged in vacuo. The STM chamber was then isolated from the rest of the system and CO was introduced at a rate of approximately 1 Torr s^{-1}. The surface was allowed to equilibrate in the high-pressure CO for 1 hour prior to imaging. Images of the surface acquired in 200 Torr of CO revealed a step morphology, similar to that seen in vacuo in the absence of CO. Enlarged views of the terraces showed no ordered structures due to CO until the pressure reached 200 Torr. Above this pressure, however, a nearly hexagonal periodic structure was formed, as shown in Fig. 2.17. This structure is only observed in the presence of CO, and has a periodicity that varies from 11 ± 1 Å in some areas of the surface to 14 ± 1 Å in other areas, with a corrugation of 0.30 ± 0.05 Å. Imaging was performed at 0.1–0.2 V bias and 0.5-1.0 nA tunneling current with a gap resistance of 200–240 MΩ. Changing either the bias or the current so that the gap became less than 200 MΩ caused the hexagonal pattern to disappear. It was recovered, however, upon restoring the initial high gap condition. Presumably, the tip gets too close to the surface at such a low gap resistance and displaces the CO. Evacuation of the CO gas to 1×10^{-4} Torr base pressure completely removed the hexagonal pattern. An important observation is that the angular orientation of the hexagonal pattern is maintained rigidly from one terrace to the next, as seen in Fig. 2.17. More details of the surface structures produced by CO are shown in Fig. 2.18. As the CO pressure is increased to

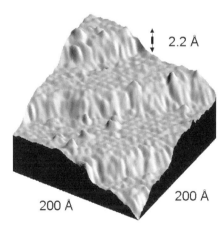

2.2 Å

200 Å

200 Å

200 Å

Fig. 2.17 3D representation of an STM image obtained in 200 Torr CO. Image size is 200 Å × 200 Å, sample bias 109 mV; tunneling current 0.52 nA. Height scale is greatly exaggerated to display corrugation on the terraces. Hexagonal arrays of maxima can be observed on each terrace due to a CO monolayer forming a Moiré-type structure. The alignment of the hexagonal array is the same in each terrace

Fig. 2.18 560 Å × 360 Å topographic STM image obtained in 150 Torr CO and 50 Torr O_2 after annealing at 183 °C; $V_s = 111$ mV, $I_T = 0.16$ nA. The inset, a 1600 Å × 2000 Å STM image, shows a monoatomic hole and mesa of Pt atoms with triangular shapes. The close-packed row directions of the Moiré pattern are aligned parallel to the step edges of the Pt structures, which are of the [110] type

750 Torr, changes in the periodicity (see line traces A and B) and phase shifts in the distribution of the hexagonal cells from one area to another can be observed. These shifts usually occur near impurities and defects (bright spots in the image).

These results indicate that the observed hexagonal structures must be due to an ordered layer of CO in equilibrium with the gas phase, which is incommensurate with the Pt(111) substrate, forming a Moiré-type structure. Small density changes in the CO overlayer that change the CO-CO distance even slightly would be considerably amplified in the Moiré periodicity, as observed in our experiments. Thus, the 11 Å periodicity observed, which corresponds to about four Pt-Pt nearest neighbor distances (2.77 Å), can include three CO molecules with an intermolecular separation of 3.67 Å and corresponds to a coverage of 0.57 ML. The 14 Å periodicity is close to five Pt distances and could include four CO molecules at 3.5 Å separation, with a corresponding coverage of 0.63 ML. These structures formed at room temperature in equilibrium with the CO gas pressure do not correspond with those observed in UHV at low temperatures. It is this structure that is likely to be relevant in catalytic processes that occur only at high pressures [31].

2.3.2.2 High-pressure NO on Rh(111)

For NO on an Rh(111) surface, it is known that under UHV conditions, a (2×2) structure with a coverage of 0.75 ML is formed either by large (~20 Langmuir) exposures to NO at 200 K, or by small (~2 Langmuir) exposures at 40 K followed by annealing to 200 K. At room temperature, it is possible to maintain this structure with a constant background of 10^{-8} Torr. X-ray photoelectron diffraction [34] and automated tensor LEED [35] studies indicate that the unit cell of this structure contains three NO molecules, one on a top site and two in threefold hollow sites. The on-top NO molecule sits 0.4 Å higher than the hollow site molecules.

Time: 0 seconds

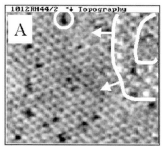

55 seconds

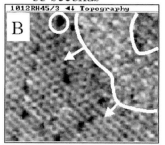

110 seconds

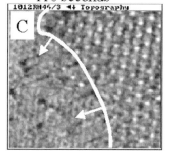

Pressure = 0.03 Torr NO

Fig. 2.19 STM images of Rh(111) taken in 0.03 Torr of NO at 25°C, showing the phase transition between a (2×2) and a (3×3) structure. The 100 Å × 100 Å images were taken at 55 second intervals (I = 440 pA, V = 99 mV). In image A, the majority of the surface shows the (2×2) structure, except for the upper right-hand corner, which shows the (3×3) structure. In images B and C, the domain boundary between the two phases moves across the image at a rate of ~20 Å per minute. Images A and B share a common defect, marked with a circle. The closepacked rows of both patterns are aligned, and the maxima separated by multiples of the Rh–Rh distance. This indicates that the bright spots correspond to NO molecules on similar sites in both the (2×2) and the (3×3) structure

In the range 10^{-8} to 0.01 Torr of NO at room temperature, the images show the same (2×2) periodicity, with only one maximum per unit cell. It is likely that the same is true for the (2×2)-3NO structure. As the NO pressure is increased in the range from 0.01 to 0.05 Torr, areas of the surface are seen to be covered with a (3×3) pattern. These areas grow over the course of a few hours until the entire surface is covered by the (3×3) structure. Figure 2.19 shows a series of 100 Å (100 Å images obtained at intervals of 55 seconds on the same area of the surface in 0.03 Torr of NO. In image A, the majority of the surface is covered with the (2×2) pattern with one corner showing a small area of the (3×3) pattern.

An island of (3×3) structure (inside the dotted line) surrounded by areas with the (2×2) structure can be seen in Fig. 2.20. Two straight lines have been drawn that separate two unit cells of the (3×3) structure or three unit cells of the (2×2) structure. The cursor profile along line A–B (shown at the bottom) reveals two interesting differences between these NO structures: (1) the corrugation of the (3×3) structure is always higher, and (2) the apparent height is also higher in the (3×3) regions. On average, the (3×3) to (2×2) corrugation ratio is 4:1 and the apparent height of the regions covered by the (3×3) structure is ~0.1 Å higher than that of the regions covered by the (2×2) structure. The higher corrugation of the (3×3) structure could be the result of the larger dimensions of the unit cell, which causes the top-site NO molecules to be farther apart and allows the tip to better follow the molecular contours. It could also be the result of a restructuring of the rhodium substrate. The

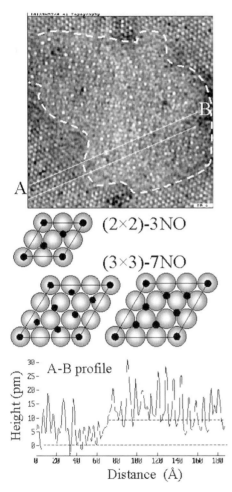

Fig. 2.20 A 200 Å × 200 Å STM image taken in 0.03 Torr NO at 25 °C, showing a (3×3) domain surrounded by the (2×2) structure. It is known that the (2×2)-3NO structure contains one top-site NO molecule and two molecules in hollow sites. Two similar models are proposed for the (3×3) structure. The first (left) consists of one top-site NO molecule and six molecules near hollow sites. In the second model (right), the near hollow site molecules are allowed to relax so that they can occupy threefold hollow sites. Both models have a 0.778 ML coverage and are consistent with the STM images. A line profile taken from line A–B on the image is shown. Note the higher corrugation in the (3×3) domain and its higher apparent height, ~0.1 Å above that of the (2×2) domain

higher elevation of the base line in the denser structure might indicate an expansion of the top layer of rhodium atoms.

While the (2×2) structure has been solved by LEED, no information is available concerning the (3×3) structure. The models shown in the middle of Fig. 2.20 are proposed on the basis of the following considerations. First, the coverage in the (3×3) structure must be only slightly higher than that in the (2×2) structure (0.75 ML) because the (2×2) structure is already very dense (only 3.16 Å separates adjacent molecules). Second, the STM images show that the rows of maxima in the two structures are parallel and are separated by multiples of the Rh–Rh distance. This indicates that the NO molecules producing the maxima occupy similar on-top sites in both structures. A (3×3) structure satisfying these two conditions and preserving the hexagonal packing of NO found in the (2×2)-3NO structure can be obtained by a rigid rotation of the NO layer by 10.9° followed by a linear compression of 1.8%. A more symmetrical structure with NO molecules in threefold hollow sites can be obtained by allowing a slight relaxation of the NO molecules inside the cell. In both of these models, the surface coverage is 0.778 ML, only 3.7% greater than the (2×2)-3NO coverage.

2.3.2.3 Tip-induced catalysis [36–38]

Since the scanning tunneling microscope (STM) can be used at atmospheric pressures and at various temperatures to provide atomic scale information on the structure of surfaces, STM is a prime technique for in situ studies of surfaces under catalytic reaction conditions. In our laboratory, we have studied several catalytic reactions of hydrocarbon clusters on Pt(111) by means of high-pressure STM in various gas environments, and different temperature and pressure regimes [38]. In this section, we review two tip-induced catalytic reactions: hydrogenation and oxidation of hydrocarbon clusters.

Preparation of hydrocarbon clusters

After the cleaned Pt(111) sample had been transferred to the reaction cell, a mixture of propylene (10%) and H_2 (90%) was admitted into the chamber with the sample at room temperature. Propylene readily adsorbs onto the surface of Pt under these conditions to form ordered structures of propylidyne ($\equiv CCH_2CH_3$), which have been studied in detailed [17, 18, 39, 40]. The propylidyne was then heated in vacuo or after backfilling the cell with 0.1 atm of CO. The decomposition of propylidyne produces fragments and clusters. Unlike propylidyne or ethylidyne, these large entities are much less mobile and can be imaged by STM. Figure 2.21 shows an example of an STM image of such hydrocarbon clusters. Only clusters produced by dehydrogenation at temperatures <650 K were found to be removable by the tip-catalyzed transfer of hydrogen and oxygen atoms.

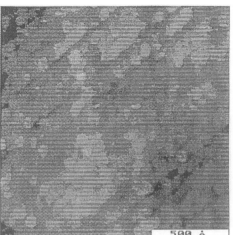

Fig. 2.21 A 1000 × 1000 Å STM image of Pt(111) after annealing the propylene-covered surface at 550 K in 0.1 atm of CO. Carbonaceous clusters are formed that uniformly cover the surface. The diameter of the clusters is about the same as the terrace width (~100 Å). Their height shows a bimodal distribution with 1 and 2 Å.

Tip-catalyzed hydrogenation of hydrocarbon clusters

After producing the carbonaceous clusters by heating in the range 470–650 K, the cell was pumped and filled again with 1 atm of propylene (1 or 10%)/H$_2$ (99 or 90%, respectively) mixtures, or with pure H$_2$, and maintained at room temperature. It was found that when the tip was "activated" by pulses of several tenths of a volt, catalytic removal of the clusters takes place as shown in Fig. 2.22. In the upper third of this image, the clusters that formed after heating the propylene layer to 550 K are seen to be aligned following the direction of the step edges (approximately along the diagonal of this picture). A pulse of 0.9 V was applied at the position marked P, which produced bumps ~14 Å high. This indicates that material

Fig. 2.22 A 3000 × 3000 Å STM image of a surface covered with carbonaceous clusters in 1 atm of an H$_2$ (90%)/propylene (10%) mixture. The clusters are visible in the upper third of the image. A voltage pulse of 0.9 V was applied to the Pt tip when it reached the position indicated, which produced a deposit of material ~15 Å high (brigth spot next to P). This pulse produced a catalytically activated tip that removed all clusters in the remaining part of the image

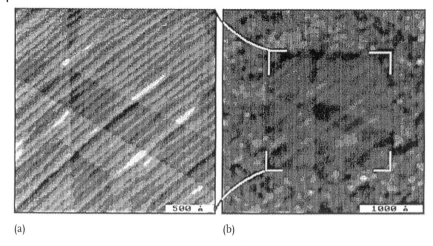

(a)   (b)

Fig. 2.23   (a) After activating the tip, a
fresh area was imaged, and all clusters
were removed so that only the step
structure characteristic of the platinum
substrate was visible. (b) Image obtained a

few minutes after the tip had
"deactivated", presumably due to
contamination. The previously scanned
area, free of clusters, is visible in the
center

had been transferred from the tip to the substrate. It can be seen that below this
line, as the tip continued to scan, the clusters have been removed. If the tip was immediately moved to a fresh area while it remained active, all the clusters were removed from the area scanned, as shown in Fig. 2.23a. Since the scanning rate was
~10^5 Å s^{-1}, the hydrogenation rate was of the order of 10^5 carbon atoms s^{-1} at 300 K,
a very fast rate indeed. The activity of the tip decayed to zero after some time (of the
order of minutes), presumably by contamination, which prevented H_2 adsorption
and dissociation at the tip. It was in this deactivated state that the image in
Fig. 2.23b was obtained. The central square corresponds to the area imaged previously, while the tip was active. Catalytic activity could readily be restored by additional pulses to remove the contaminant material, as shown in the sequence of images in Fig. 2.24. The first image (Fig. 2.24a) corresponds to a large area covered by
clusters (seen more clearly in the expanded area, Fig. 2.24b), together with a couple
of large features in the upper right-hand corner. These two features are identical in
shape, indicating that they correspond to a single object imaged twice by a double
tip apex. The tip was catalytically inactive during the acquisition of this image. Immediately afterwards, the tip was pulsed twice near the center, at the position
marked by stars in Fig. 2.24a. The two bumps in the center of Fig. 2.24c (marked
p1), are the result of the application of these pulses. The activated tip obtained in
this way was used to image the same area. As shown in Fig. 2.24c, all the clusters
were removed in the top part of the image (see the expanded area in Fig. 2.24d).
The large feature was also thinned out (its height was halved after passage of the ac-

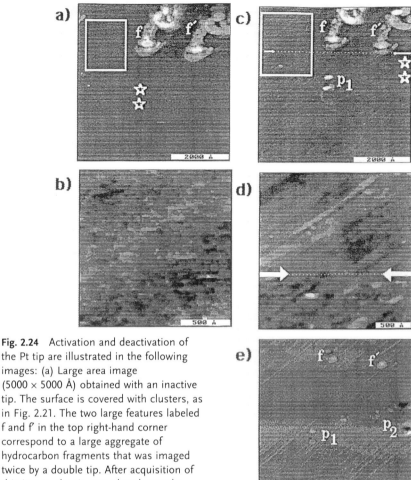

Fig. 2.24 Activation and deactivation of the Pt tip are illustrated in the following images: (a) Large area image (5000 × 5000 Å) obtained with an inactive tip. The surface is covered with clusters, as in Fig. 2.21. The two large features labeled f and f' in the top right-hand corner correspond to a large aggregate of hydrocarbon fragments that was imaged twice by a double tip. After acquisition of this image, the tip was placed near the center and two pulses of 0.9 V were applied. (b) Enlarged image of the square in (a), showing the clusters. (c) While the activated tip, this new image was acquired. The clusters in the top third were all removed. The large feature f is still visible although its height has been halved. At the line scan position marked by dots and arrows the tip became deactivated, presumably due to contamination, and no cluster removal occurred thereafter. A magnified image of the square is shown in (d). p_1 marks the position of the bumps left by the previous activating pulses. At the end of the image two new pulses were applied at the position marked by stars, on the other side. (e) Image obtained after the second pair of pulses. This time the tip remained active over the entire image and all the clusters were removed. Feature f was also removed, except for a remaining unreactive core; p_2 indicate the marks left by the second activating pair of pulses

tive tip), indicating that it was made of hydrocarbon material that was also reacted away by the tip. In this experiment, the tip became deactivated at approximately the scan line marked by the arrows and the remainder of the image exhibits the intact original clusters. Another couple of pulses at the positions marked by stars in Fig. 2.24c reactivated the tip once more. In the following scan shown in Fig. 2.24e, all clusters were removed. The large features f and f' are seen to have been almost completely removed except for a residual core. The bumps from the previous two pulses can clearly be seen at the position marked p_2.

Tip-catalyzed oxidation of hydrocarbon clusters

In addition to hydrogenation, tip-catalyzed oxidation of hydrocarbon fragments at 300 K has also been observed. To carry out the reaction, the cell was filled with 1 atm of pure oxygen. A similar procedure as in the previous case was followed to activate the tip, using voltage pulses to remove contamination. Fig. 2.25 shows an example of these results. The top image shows the surface before tip activation, whereas the lower image was taken on the same area after activation. Note that not all the hydrocarbon clusters were removed from the scanned region. It is difficult

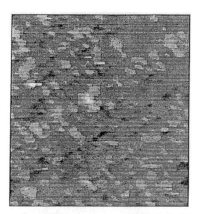

Fig. 2.25 1000 × 1000 Å images illustrating the tip-catalyzed removal of clusters in 1 atm of pure oxygen. The top image was obtained before activation; the lower image was obtained after activation by the application of a 1 V pulse

to quantitatively compare the rates of carbon removal by oxidation and hydrogenation because of several other variables that influence the rates. It appears, however, that the platinum tips remain active and free of contamination longer in oxygen than in hydrogen or in hydrogen/propylene mixtures.

2.4
Challenges and Future Directions

The development of techniques that can monitor so-called buried interfaces, solid–high pressure gas, solid-liquid, and solid–solid interfaces at the molecular level are leading to studies of new surface phenomena and new surface materials. Nanoparticles in the 2–100 nm range are being synthesized and their atomic and electronic structures are being studied as a function of size. Their unique size-dependent chemical and electronic properties are likely to lead to the discovery of highly selective catalysts, absorbers, and new electroluminescent and sensor materials. Studies of molecules at solid–liquid interfaces in the presence of external potentials allow unique control over their orientation and reactivity. Surface electrochemistry that monitors the structure and bonding of adsorbed molecules, in addition to the change of current as a function of applied potential, will lead to the discovery of more efficient fuel cells and chemical sensors.

Biopolymer surfaces, such as skin, bone, and man-made implant polymers, will be studied in the presence of protein solutions to uncover the causes of bio-incompatibility. Better temporal and spatial resolution of the instrumentation for studying surfaces should permit studies of dynamic processes at surfaces, diffusion phenomena, the kinetics of surface reactions, and molecular rearrangements. Surface materials will remain at the frontier of solid-state and materials chemistry for years to come.

Acknowledgement

This work was supported by The Director, Office of Energy Research, Office of Basic Energy Sciences, Materials Science Division, of the US Department of Energy.

References

1 Blakely, D. W.; Kozak, E. I.; Sexton, B. A.; Somorjai, G. A., *J. Vac. Sci. Technol.* **15**, 1091 (1976).

2 Du, Q.; Superfine, R.; Freysz, E.; Shen, Y. R., *Phys. Rev. Lett.* **70**, 2313 (1993).

3 Shen, Y. R., *Nature* **337**, 519 (1989).

4 Shen, Y. R., *"The Principles of Nonlinear Optics"*, John Wiley Inc., New York, 1984.

5 Guyot-Sionnest, P.; Hunt, J. H.; Shen, Y. R., *Phys. Rev. Lett.* **59**, 1597 (1987).

6 Bain, C. D., *J. Chem. Soc., Faraday Trans.* **91**, 1281 (1995).

7 Horiuti, I.; Polanyi, M., *Trans. Faraday Soc.* **30**, 1164 (1934).

8 Cassuto, A.; Kiss, J.; White, J., *Surf. Sci.* **255**, 289 (1991).

9 Ibach, H.; Lehwald, S., *Surf. Sci.* **117**, 685 (1982).

10 Cremer, P.; Stanners, C.; Niemantsverdriet, J.; Shen, Y.; Somorjai, G., *Surf. Sci.* **328**, 111 (1993).

11 Land, T.; Michely, T.; Behm, R.; Hemminger, J.; Comsa, G., *J. Chem. Phys.* **97**, 6774 (1992).

12 Davis, S.; Zaera, F.; Gordon, B.; Somorjai, G., *J. Catal.* **92**, 250 (1985).

13 Beebe, T.; Yates, J., *J. Am. Chem. Soc.* **108**, 663 (1986).

14 Mohsin, S.; Trenary, M.; Robota, H., *J. Phys. Chem.* **92**, 5229 (1988).

15 Cremer, P. S.; Su, X.; Shen, Y. R.; Somorjai, G. A., *J. Am. Chem. Soc.* **118**, 2942 (1996).

16 Schlatter, J.; Boudart, M., *J. Catal.* **24**, 482 (1972).

17 Avery, N.; Sheppard, N., *Proc. Roy. Soc. London* **A405**, 1 (1986).

18 Koestner, R.; Frost, J.; Stair, P.; Van Hove, M.; Somorjai, G., *Surf. Sci.* **116**, 85 (1982).

19 Lok, L.; Gaidai, N.; Kiperman, S., *Kinet. Katal.* **32**, 6, 1406 (1991).

20 Otero-Schipper, P.; Wachter, W.; Butt, J.; Burwell, R.; Cohen, J., *J. Catal.* **50**, 494 (1977).

21 Shahid, G.; Sheppard, N., *Spectrochim. Acta* **46A**, 6, 999 (1990).

22 Bussell, M. E.; Henn, F. C.; Campbell, C. T., *J. Phys. Chem.* **96**, 5965 (1992).

23 Rodriguez, J. A.; Campbell, C. T., *J. Phys. Chem.* **93**, 826 (1989).

24 Lamont, C. L. A.; Borbach, M.; Martin, R.; Gardner, P.; Jones, T. S.; Conrad, H.; Bradshaw, A. M., *Surf. Sci.* **374**, 215 (1997).

25 Land, D. P.; Erley, W.; Ibach, H., *Surf. Sci.* **289**, 773 (1993).

26 Martin, R.; Gardner, P.; Tushaus, M.; Bonev, C. H.; Bradshaw, A. M.; Jones, T. S., *J. Electron Spectrosc. Relat. Phenom.* **54/55**, 773 (1990).

27 Su, X.; Kung, K.; Lahtinen, J.; Shen, Y. R.; Somorjai, G. A., *Catal. Lett.* **54**, 9 (1998).

28 Yoshinobu, J.; Kawai, M., *J. Chem. Phys.* **103**, 3220 (1995).

29 Campbell, C. T.; Ertl, G.; Kuipers, H.; Segner, J., *J. Chem. Phys.* **73**, 5862 (1980).

30 Yoshinobu, J.; Kawai, M., *J. Chem. Phys.* **103**, 3220 (1995).

31 Su, X.; Cremer, P. S.; Shen, Y. R.; Somorjai, G. A., *J. Am. Chem. Soc.* **119**, 3994 (1997).

32 Su, X.; Cremer, P. S.; Shen, Y. R.; Somorjai, G. A., *Phys. Rev. Lett.* **77**, 3858 (1996).

33 Jensen, J. A.; Rider, K. B.; Salmeron, M.; Somorjai, G. A., *Phys. Rev. Lett.* **80**, 1228 (1998).

34 Kim, Y. J.; Thevuthasan, S.; Herman, G. S.; Peden; Eden C. H. F. et al., *Surf. Sci.* **359**, 269 (1996).

35 Zasada, I.; Van Hove, M. A.; Somorjai, G. A., *Surf. Sci.* **418**, L89 (1998).

36 McIntyre, B. J.; Salmeron, M.; Somorjai, G. A., *Science* **265**, 1415 (1994).

37 McIntyre, B. J.; Salmeron, M.; Somorjai, G. A., *Catal. Lett.* **39**, 5 (1996).

38 McIntyre, B. J.; Salmeron, M.; Somorjai, G. A., *J. Catal.* **164**, 184 (1976).

39 Koestner, R.; Van Hove, M.; Somorjai, G., *J. Phys. Chem.* **87**, 203 (1983).

40 Salmeron, M.; Somorjai, G. A, *J. Phys. Chem.* **86**, 341 (1982).

3
In Situ NMR

James F. Haw

3.1
Introduction

Nuclear magnetic resonance is the most widely used method by which chemists elucidate molecular structure and it is widely used to study dynamics and reactivity. Solution-state NMR of ^{1}H, ^{13}C, ^{31}P, and other nuclei has greatly contributed to our understanding of the mechanisms of homogeneous catalysis, and solution methods have been extended to permit in situ studies at high temperatures and pressures. Solid-state NMR is widely used to characterize the structures of solid catalysts; for example, over 600 papers since 1989 have both "NMR" and "zeolite(s)" in their titles. These papers include ^{29}Si and ^{27}Al studies of framework atoms [1–4], ^{1}H studies of acid sites [5–7], ^{13}C, ^{15}N, and ^{31}P studies of probe molecules [8–14], and ^{129}Xe studies of adsorbed Xe [15], a subtle probe of structure and topology. Breakthroughs in averaging methods such as MQ-MAS [16–19] have led to a revolution in solid-state NMR of half-integer quadrupolar nuclei, and many ^{27}Al sites on catalysts that were previously difficult to characterize are now resolvable by suitable NMR experiments. Even ^{17}O NMR has become chemically useful, although isotopic enrichment is required [20].

The most inclusive definition of in situ NMR [21–26] in heterogeneous catalysis research is the use of nuclear magnetic resonance to better understand a reaction on a solid catalyst. Usually, this has been taken to mean NMR measurements performed on starting materials, intermediates, and/or products adsorbed onto the catalyst. For example, most catalytic reactions involve one or more carbon-containing reactant, and ^{13}C NMR studies of adsorbates on solid catalysts are very common. In situ NMR measurements may also probe the structure of the catalyst as it is modified under reaction conditions. Thus, even an ^{27}Al NMR study of zeolite dealumination by high-temperature steaming could be said to be an in situ investigation as the modification of the catalyst under reaction conditions can have a distinct bearing on reaction mechanism. In a similar vein, various NMR experiments have been used to follow the formation of zeolites and related materials in situ as they are synthesized under hydrothermal conditions. This too is a form of in situ NMR applied to catalysis research.

A far more restrictive definition of in situ NMR is one in which the catalytic reaction must take place inside the NMR receiver coil, and the spectra are acquired at high temperature during the conversion of reactants to products. This purist view of in situ spectroscopy drove the development of in situ NMR for some time and led to a number of technological advances in NMR probes, sample rotors, and related NMR or sample preparation hardware. These devices also revealed important information about heterogeneous catalysis, but the purist approach to in situ NMR has had to contend with the essential dissimilarities between catalytic flow reactors and the best solid-state NMR sample containers. More fundamentally, the highest quality NMR measurement usually requires a relatively low sample temperature and a lengthy observation time, while catalytic activity requires a high temperature and selectivity is favored by a short contact time. There seems to be no engineering solution to the latter problems, at least until a broadly applicable method for polarizing NMR transitions far beyond their thermal values is discovered. Thus, much of the recent development of in situ NMR has relaxed the constraints on whether the spectroscopy is in situ in the purest sense and has instead stressed the use of NMR measurements to learn about heterogeneous catalysis by whatever means are necessary. This more pragmatic approach has allowed NMR to make important contributions to the solution of an important and long-standing mechanistic problem in heterogeneous catalysis, that of methanol conversion to hydrocarbons.

This chapter will briefly survey some of the more important experimental methods for in situ NMR studies of catalysis. Most of these methods are applicable to thermal reactions, and they are easily compared with each other in terms of strengths and limitations. One exception is an in situ NMR method for the study of photocatalytic reactions developed by Raftery and co-workers [27–30]. This method is presented first, even though it is a recent development, so as not to break the treatment of in situ methods developed for thermal reactions and their application to a common catalytic application. It is argued that all of the methods of in situ NMR are useful and that some of the less frequently used methods could enjoy a renaissance if applied to catalytic problems that fit their particular strengths. The chapter then focuses on the application of in situ NMR in studying the mechanism by which methanol is converted to hydrocarbons on zeolites and other microporous solid acids. Every thermal method of in situ NMR has been applied to methanol conversion chemistry, and the number of papers on this problem easily exceeds that for any other application with the possible exception of the chemistry of carbenium ions on solid acids [31–34].

3.2
Methods of In Situ NMR

3.2.1
General Considerations

An NMR measurement of a solid catalyst could focus on molecules in the gas phase, on rigid or nearly rigid species on or of the catalyst itself, or on species of intermediate mobility. Each of these poses a different challenge to NMR observation. Gas-phase species are dilute, and under flow conditions they might have a short residence time in the receiver coil region. A rigid molecule will have appreciable static dipolar interactions between spins that may require strong decoupling fields, and the orientation dependence of the chemical shift will be manifested as broad powder pattern line shapes unless sample rotation, most commonly magic-

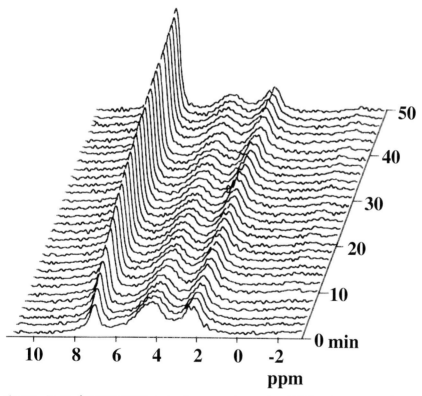

Fig. 3.1 In situ ¹H MAS NMR study of H/D isotope exchange between the acid site of zeolite HZSM-5 and [D₆]benzene. The sample was prepared by adsorption at reduced temperature, and the temperature was jumped to 333 K in the NMR probe. ¹H MAS NMR spectra measured over time showed a decrease in the signal due to the acid site at $\delta = 4.3$ and an increase in the signal due to protons on benzene ($\delta \approx 7$).

angle spinning (MAS), is applied. Strong ^1H–^1H dipolar couplings in rigid solids usually limit the application of in situ ^1H NMR to cases of either rare (chemically or isotopically dilute) protons or cases of appreciable molecular motion. The Brønsted acid sites of zeolites are far enough apart that ^1H chemical shift information for these sites can be obtained using MAS as the only line-narrowing technique. Resolved ^1H spectra can also be obtained for mobile organic species on catalysts. For example, methylbenzenes adsorbed on large-pore zeolites may show resolved methyl and aromatic ^1H signals.

Figure 3.1 shows a successful application of ^1H MAS NMR to an interesting reaction on a solid catalyst [35]. [D_6]Benzene was adsorbed on the solid acid zeolite HZSM-5 at low temperature, the sample was transferred to the NMR probe, and the temperature was then jumped to a higher value. ^1H NMR spectra collected over time yielded the rate of H/D exchange between the zeolite acid site and the aromatic ring, the most simple example of electrophilic aromatic substitution. This experiment was performed several times at various temperatures, and the apparent activation energy evaluated from the rate data was consistent with theoretical modeling of the reactants and transition state. The application in Fig. 3.1 was successful in part because of the simple spectra and large ^1H chemical shift differences between the acid site and adsorbed benzene. In general, ^1H NMR studies of organic species on a solid catalyst should be interpreted with caution, as the more catalytically relevant species could be under-represented in the spectrum.

A clear majority of the papers published on the use of NMR to study reactions on solid catalysts have used ^{13}C. This follows from the primacy of organic reactions in heterogeneous catalysis, the large chemical shift range for ^{13}C, and the manageable costs of isotopic enrichment. Well-resolved ^{13}C spectra of rigid solids require high-power ^1H decoupling and usually averaging of the ^{13}C chemical shift anisotropy by sample rotation. Most of the results presented in this chapter are based on ^{13}C spectra.

3.2.2
In Situ NMR of Photocatalysis

Figure 3.2 shows the important features of Raftery's photocatalysis experiment [27–30]. A key innovation was coating optical microfibers with small TiO_2 catalyst particles to provide enough irradiated catalyst surface to permit NMR detection of the photoproducts. Figure 3.2 shows how the microfibers are packed into a glass ampoule, which also holds the reactant, ethanol, and maintains a controlled atmosphere. This glass ampoule is inserted into a plastic sleeve, which is machined as a gas-turbine rotor. The whole assembly is placed into the spinning module that provides the stator for rapid sample rotation on compressed gas streams, as well as housing the radio-frequency (RF) coils and a quartz rod that guides UV light to the sample region probed by NMR. In this, as in many other in situ NMR experiments,

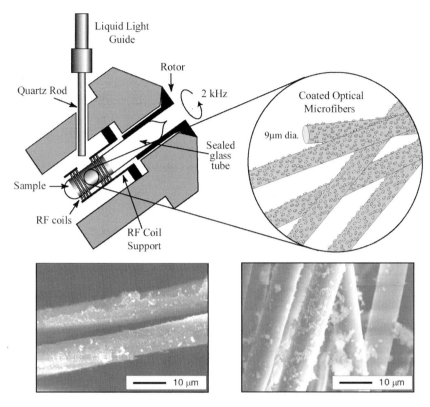

Fig. 3.2 Schematic diagram showing an in situ photocatalysis NMR probe and coated optical fibers. SEM images are also shown for TiO₂ and zeolite-coated 9 μm optical microfiber photocatalysts.

the sample is rotated at several thousand times per second; this rapid rotation averages the orientation dependence of the chemical shift and can provide narrow NMR signals when the rotor is inclined at the magic angle (54.7°) relative to the applied magnetic field. Figure 3.3 shows ^{13}C MAS NMR spectra acquired during UV irradiation of ethanol on TiO_2 using the apparatus described above. Signals due to the reactant gave way to those of product species during irradiation.

3.2.3
In Situ NMR of Thermal Reactions in Sealed MAS Rotors

Most thermal in situ NMR applications to date have used batch reactor conditions with the reactant and catalyst sealed in a sample rotor with no provision for removing products or admitting additional reactants. There are a number of variations

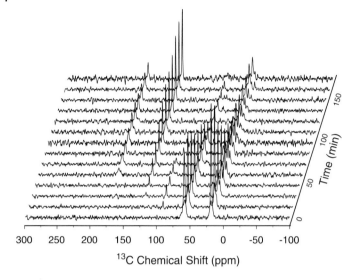

300   250   200   150   100   50   0   -50   -100

^{13}C Chemical Shift (ppm)

Fig. 3.3 In situ ^{13}C MAS spectra showing the time course of the photocatalytic oxidation of ethanol over a TiO$_2$ nanoparticle catalyst supported on optical microfibers. Ethanol (δ = 17 and 57) is converted to acetal (δ = 15, 23, 62, and 94), acetic acid (δ = 20 and 176), and CO$_2$ (δ = 125).

on this theme. Most simply, the reactant is adsorbed onto the catalyst in a glass ampoule, which is then flame-sealed and placed in a sample rotor for the measurement of an NMR spectrum [36–38]. The ampoule can be heated in an oven to progressively higher temperatures between NMR acquisitions at room temperature. This approach has an obvious advantage in that a standard NMR instrument is used in a completely standard way with no compromises other than possibly a tolerable reduction in sample payload as a result of the glass ampoule. This approach is sometimes called controlled-atmosphere NMR [39–41].

A more purely in situ approach to studies in sealed MAS rotors is to heat the sample in a variable-temperature (VT) MAS NMR probe and to acquire spectra at reaction temperature. This approach became possible with the availability of suitable VT-MAS NMR probes in the 1980's and the development of procedures for transferring catalyst/reactant samples from vacuum lines to MAS rotors. Figure 3.4 shows a diagram of a shallow-bed CAVERN device, a glass apparatus for catalyst activation and reagent adsorption that also transfers the sample into an MAS rotor [25, 42, 43]. The adsorption step in a CAVERN can be carried out at reduced temperatures; the first application was to the study of propene oligomerization in zeolite HY, and it required that the sample remained at cryogenic temperatures from adsorption to acquisition of the first MAS spectra. The probe temperature

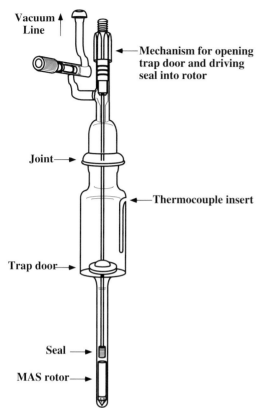

Vacuum Line

Mechanism for opening trap door and driving seal into rotor

Joint

Thermocouple insert

Trap door

Seal

MAS rotor

Fig. 3.4 CAVERN device used for preparing samples of solid acids and reactants for in situ NMR study. A shallow bed of solid acid is spread over the trapdoor, and the device is assembled and attached to the vacuum line. After evacuating and heating to generate the active form of the acid, cryogen is brought up to the level of the catalyst (if needed) and a controlled amount of reactant is adsorbed. Rotation of the shaft first raises the trapdoor, dropping the sample into the MAS rotor, then drives the seal into place.

was then slowly raised and spectra were acquired as propene reacted, through what turned out to be an unexpected pathway [44] (see Chapter 1).

Reliable commercial MAS NMR probes are available that operate at temperatures as high as 523 K using hot nitrogen gas to bathe the sample rotor. While some catalytic processes do indeed operate at this temperature, the in situ NMR applications reported in the literature have generally used 523 K to approximate conditions that actually take place industrially at higher temperatures, with slower kinetics easing NMR spectral acquisition. The ^{13}C MAS NMR spectra in Fig. 3.5 provide an example of this approach to in situ NMR for the case of methanol conversion to hydrocarbons on the zeolite solid acid HZSM-5 [45]. [^{13}C]Methanol was adsorbed onto the activated zeolite, which was then sealed into an MAS rotor using a CAVERN device. Hot gas was used to heat the sample in the NMR probe from room temperature to 250°C in several minutes. ^{13}C MAS NMR spectra were acquired at this temperature at various times between 5 and 75 minutes. Briefly, these spectra show the conversion of methanol (δ = 50), first to dimethyl ether

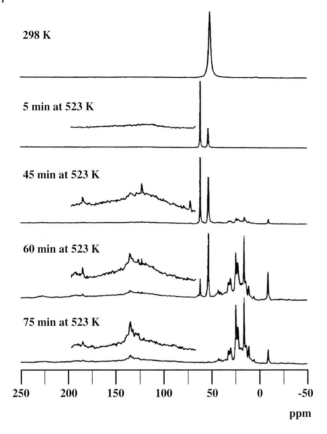

Fig. 3.5 In situ ^{13}C MAS NMR study of the reactions of [^{13}C]methanol in a sealed variable-temperature rotor. This sample was prepared using a CAVERN. The sample temperature was increased to 523 K over a period of several minutes, and then NMR spectra were acquired over the course of 75 min as methanol was converted first to dimethyl ether and then to hydrocarbon products.

($\delta = 60$), and then to a plethora of hydrocarbon products. This result was at or near the state-of-the-art in 1992, but it actually revealed very little about reaction mechanism.

3.2.4
Sealed Rotors with Transient Heating

It is possible to build conventional variable-temperature MAS NMR probes that operate at even higher temperatures [46], but their reliability is not high, and the

NMR sensitivity is reduced due to low Boltzmann factors and high receiver coil noise. High-power infrared lasers can be used to heat NMR samples to very high temperatures very quickly. Freude and co-workers [47-49] and the author's group [50] have described transient in situ NMR experiments based on IR laser heating. There have been a few variations on this experiment, but most simply a sealed sample of a catalyst and reactant spinning in an MAS probe is subjected to a short (ca. 5 to 30 s) period of intense IR heating, during which its temperature increases up to several hundred degrees. NMR spectra are then acquired either at the reaction temperature or more commonly after the rapid cooling that occurs after laser irradiation is terminated. A series of NMR spectra are acquired as a function of total heating period or some other experimental variable to map a progression from reactants to products.

Figure 3.6 shows a diagram of an MAS NMR experiment with laser heating. Note that the IR laser beam must enter either the top (or more conveniently) the bottom of the magnet and then be focussed onto a narrow spot so as to miss the RF coil and bathe the sample region as the rotor turns. Generally, the laser source is some distance from the magnet and the beam must be reflected by 90° to enter the vertical magnet bore. Several optical components are required for collimation and focussing. The appeal of this experiment is that the attainment of very high temperatures for short periods (i.e. seconds or tens of seconds) is much closer to true in situ catalysis conditions than heating at lower temperatures for several minutes or longer. However, using an IR laser to transiently heat an MAS NMR rotor is instrumentally challenging. There is no convenient way of measuring the temperature inside a rotor spinning at several thousand rotations per second on gas bearings. The best that one can do is to calibrate the experiment using an NMR chemical shift thermometer such as the ^{207}Pb chemical shift of solid lead nitrate, and assume the same time-temperature profile for the experimental sample given identical laser power [50]. IR laser heating also results in significant temperature gradients across the catalyst bed, and these evolve with time. Finally, working with a high-power IR laser is very dangerous, especially during optical alignment.

An alternative approach to transient heating involves the use of MAS rotors that are thinly coated with Pt metal, as is also illustrated in Fig. 3.6. The Pt coating is inductively heated by high-power radio-frequency irradiation, which is conveniently applied by switching on the NMR probe's ^{1}H decoupler channel for seconds or tens of seconds. This approach has problems in common with the IR laser experiment; the temperature must be calibrated against an external standard, and it can be very inhomogeneous, especially during a temperature jump. Nevertheless, inductive heating using Pt-coated MAS rotors is a safe approach to in situ NMR that does not require much instrumental modification. Figure 3.7 shows the application of inductive heating to methanol conversion on zeolite HZSM-5 [50]. In this case, single-transient (i.e. one scan) ^{1}H NMR spectra were taken at three-second

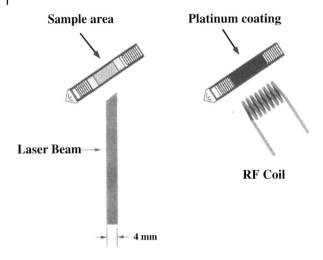

Laser Heating   **RF Heating**

Fig. 3.6 Left: Experimental geometry for MAS NMR with laser heating. The split solenoid radio-frequency coil with a gap to admit the laser beam is shown off-set for clarity. The beam trace on the MAS rotor is shaded. Quartz MAS rotors were used; these absorb the 10.6 µm radiation but conduct the heat to the sample in several seconds. **Right:** Experimental geometry for MAS NMR with inductive heating. The radio-frequency coil that was used for transmitting and receiving NMR signals and for inductive heating is shown off-set from the platinum-plated zirconia rotor for clarity. This exploded view also illustrates the thermal bulkheads and grooved Kel-F seals. The platinum coating was typically 25% longer than the RF coil, and the sample was limited to the central 75% of the length of the RF coil for a nominal payload of 75 µL.

intervals during application of an 18 W heating pulse 48 s in duration. Calibration showed that the sample reached a steady-state temperature of 350 °C after ca. 30 s of irradiation.

The broad ^{1}H MAS spectra in Fig. 3.7 are frankly of little mechanistic value; one can see a narrowing of the methanol line during the temperature jump (first 30 s) followed by the formation of additional peaks as the reaction occurred, but the resolution permits little in the way of assignments. Figure 3.7 suggests that although it is technically possible to carry out a "true" in situ NMR experiment on time scales comparable to the contact times in industrial catalysis, the quality of the spectroscopy can be severely compromised. Of course, the experiments in Figs. 3.5 and

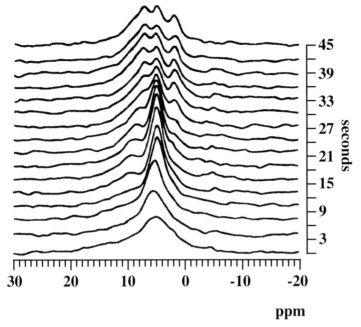

Fig. 3.7 Single-shot 199.6 MHz ^{1}H MAS spectra of zeolite HZSM-5 with a 3.5 mmol/g loading of [^{13}C]methanol acquired at 3 s intervals during a 48 s, 18 W inductive heating pulse. A final mean temperature of 623 K was reached in 30–35 s.

3.7 did not provide for reagent flow, and overcoming this limitation motivated much in situ NMR development work.

3.2.5
In Situ MAS Flow Probes

Reimer and co-workers developed one of the first in situ NMR probes, which was essentially a conventional bench-top flow reactor design assembled inside a high-temperature NMR probe [51]. The catalyst bed was stationary in this probe, and no attempt was made to average the orientation dependence of the chemical shift. Thus, the NMR signals were very broad, and applications were restricted to reactions that involved only a few species with very distinct chemical shift tensors. Nevertheless, the Reimer flow probe was a true catalytic reactor, and it motivated several attempts at improvement by maintaining reagent flow while providing for averaging of the orientation dependence of the chemical shift. The groups of Hunger [52–55], Munson [56, 57], and the author [58] have each described high-temperature MAS probes whereby gases are introduced into and removed from a catalyst-packed MAS rotor through tubes penetrating along the spinning axis. Figure 3.8 shows a diagram of the Munson design. Since the flow tubes are stationary in all of these designs and the rotor is spinning at several thousand times per second, contact cannot be made between the rotor and the flow tubes, and the reactor is not gas-tight. Rather, a gas stream containing reactants is passed into one end of the rotor where it contacts the catalyst bed. Because of centripetal forces, the catalyst bed is packed as a concentric tube against the rotor wall, and there is an open pathway through the center of the rotor. The gas stream enters the rotor and equilibrates with the catalyst bed, but it is not actually forced through the bed by a pressure gradient, as in a true flow reactor. Some of the gas exiting the rotor is captured in a second tube at the other end, and this stream can be analyzed by gas chromatography.

Flow MAS probes are inspired pieces of instrument development, but like other true in situ NMR experiments they suffer from NMR sensitivity loss at high temperatures, and material constraints complicate efforts to reliably extend MAS NMR to the higher temperatures actually used for industrial catalytic processes. Perhaps most importantly, since gas-tight seals between the rotor and the flow

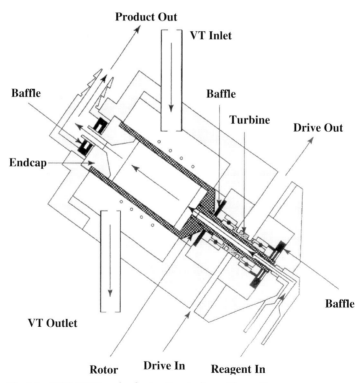

Fig. 3.8 MAS NMR probe for in situ studies with continuous introduction of reactants and removal of products. The design developed by Prof. Munson is shown.

tubes are impossible, mass transport between the inner and outer surfaces of the annular catalyst bed relies on diffusion, and the space velocity across the catalyst bed is not uniform. Most flow MAS applications have sought to understand the mechanisms of various organic reactions, but the technique might be better applied to determine catalyst structure under equilibrium conditions. For example, it could be used to study variations in catalyst composition or active site structure as a function of temperature or water partial pressure. A potentially important application of flow MAS NMR involves the introduction of optically polarized xenon gas [59], which must be polarized outside of the sample and can then be adsorbed using a flow MAS rotor. If an efficient, broadly applicable method for transferring polarization from xenon to surface species is developed, flow MAS NMR will become a very widely used experiment for the study of surface structure and reaction mechanisms.

3.2.6
Magic-Angle Hopping Flow Probes

The groups of Maciel [60] and Raftery [61] have independently described in situ flow probes based on the magic-angle hopping experiment. Magic-angle spinning uses rapid and continuous sample rotation to average the orientation dependence of the chemical shift. Maciel and co-workers showed that averaging could also be achieved using a series of discrete hops to visit three angular orientations relative to the applied magnetic field [62]. A two-dimensional NMR data set is generated by a number of hopping experiments, and an isotropic NMR spectrum of the sample can then be obtained by projection. The advantages of a hopping experiment for in situ spectroscopy are twofold. Since a hopping experiment can be performed without a complete rotation of the sample container, gas-tight seals can be achieved by using flexible tubing. Secondly, the material constraints on a hopping experiment are less severe than for high-speed spinning, and it should be possible to realize a mechanically robust hopper probe that functions at very high temperatures. Of course, the Boltzmann factor is a concern in any high-temperature NMR experiment, and the need to acquire a 2D data set is an additional challenge to sensitivity. However, the hopper experiment deserves more attention as a means of achieving a true in situ NMR measurement, especially for steady-state conditions, under which a long acquisition time is not problematic.

3.2.7
In Situ NMR using Quench Reactors

In situ NMR experiments using sealed sample tubes have revealed quite a lot about chemistry in zeolites and have led to a major revision of ideas about acid strength and the nature of carbenium ion intermediates in such media. Flow MAS, hopper experiments, and transient heating are all better approximations of a "true" in situ NMR experiment, but several years of development efforts using such experiments have not significantly improved our understanding of catalytic reaction mechanisms. A fundamentally important limitation of earlier in situ NMR experiments is that they are not readily able to probe reactions after the transient introduction of reactants to the catalyst bed. Real catalytic reactions are run with contact times ranging from a second to several tens of seconds. In a mechanistic investigation, even shorter contact times may be advantageous.

In 1998, we reported a different approach to in situ NMR – the pulse-quench reactor [63, 64]. Figure 3.9 shows a diagram of this device. It has all of the characteristics of a standard research microreactor, including provision for continuous and/or pulsed introduction of reactants and GC or GC-MS analysis of volatile products that exit the reactor. The most significant feature of the quench reactor is a provision for rapidly switching the gas stream passed over the catalyst bed to cryogenically cooled nitrogen so as to very rapidly reduce the catalyst temperature. Instrumentally, this is achieved through the use of high speed, pneumatically actuated valves and computer control. The temperature of the catalyst bed is measured by a high-speed thermocouple in contact with zeolite pellets at the downstream end of the catalyst bed. This is illustrated in Fig. 3.10, which shows time-temperature profiles from three closely related experiments in which ethylene was pulsed onto zeolite HZSM-5 catalyst beds at various temperatures [65]. The thermocouple, placed near the front of the reactor, responded to the temperature increase arising from the heat of reaction of the olefin on the solid acid. In each case, the command for a thermal quench was given 4 s after the ethylene pulse. The data in

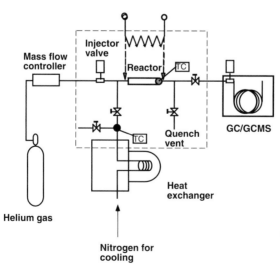

Fig. 3.9 Schematic of the pulse-quench catalytic reactor used to prepare samples under standard flow reactor conditions for NMR study. The dashed box denotes the part of the reactor that is removed to a glove box prior to transfer of the quenched catalyst sample to an NMR rotor. All valves and the heating element are under computer control.

Fig. 3.10 show that the catalyst temperature sometimes remained unchanged for a few tenths of a second due to the timing of the gas flows, but once the quench gas reached the catalyst bed its temperature plunged very rapidly. With very careful adjustment of the gas flow, we have achieved a time resolution in pulse-quench experiments as short as 200 ms.

There are a number of variations on quench reactor experiments that differ in the number of pulses of the same or diverse reagents, the use of syringe pumps instead of pulse valves to probe steady-state conditions, and so on. Most simply, a series of samples is prepared by first preparing catalyst beds at some fixed temperature, pulsing a quantity of reactant, and then allowing the composition of the catalyst bed to evolve for a variable time under continuous inert carrier gas flow. The chemistry is then interrupted by a thermal quench, which ideally fossilizes the catalyst bed such that its composition is closely related to that immediately prior to the pulse. Figure 3.11 shows the timing diagram for this single-pulse experiment as well as that for a double-pulse experiment. Double-pulse experiments are useful for probing kinetic induction periods as well as carbon isotope exchange. For example, a pulse of [^{12}C]methanol might be delivered in a first pulse followed sometime later by a pulse of [^{13}C]methanol. NMR of the quenched catalyst then reveals those structures that preferentially exchanged labels, while GC-MS analysis of gas samples collected following each pulse and immediately prior to quench provides complementary information about the isotopic composition of volatile products.

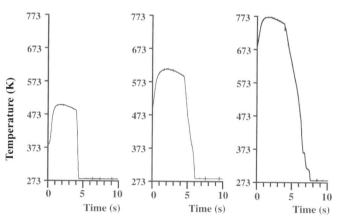

Fig. 3.10 Selected pulse-quench temperature profiles of the reactions of [^{13}C$_2$]ethylene (17.6 mg ethylene per pulse, corresponding to 1.9 molecules per acid site) on zeolite HZSM-5 at various initial temperatures. The thermocouple was placed near the front of the catalyst bed. Note the thermal transients from the heats of adsorption and reaction and the thermal quenches following 4 s of reaction.

Single pulse experiment

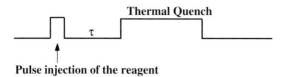

Pulse injection of the reagent

Double pulse experiment

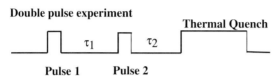

Pulse 1   Pulse 2

Fig. 3.11 Conceptual timing diagrams for pulse-quench NMR studies with either one or two reagent pulses. Product gas samples may be collected for G-MS analysis at any point. A single NMR observation is made for each catalyst bed following a temperature quench. A series of experiments varying, for example, the time between a pulse and the quench, would form a pulse-quench NMR study.

Following a quench, the reactor is sealed and moved into a glove box, where it is opened under an inert atmosphere. The entire catalyst bed is transferred to an MAS rotor, which is then sealed. Solid-state NMR spectra are then acquired at leisure, usually at room temperature. Figures 3.12 and 3.13 show complementary examples of the pulse-quench approach to in situ NMR [65]. Ethylene is one of the primary products of the conversion of methanol to hydrocarbons, and it is important to understand how this olefin reacts in the catalyst to form other species. The pulse-quench results in Fig. 3.12 show the evolution of organic products on zeolite HZSM-5 catalyst beds at a fixed temperature, 623 K, at times ranging from 0.5 s to 16 s following admission of a pulse of $[^{13}C_2]$ethylene. For the experiments in Fig. 3.13, the reaction time was kept fixed at 4 s, but the catalyst bed temperature was varied over a wide range, from 323 K to 773 K. These results show that a major product that accumulates on the catalyst is the 1,3-dimethylcyclopentenyl carbenium ion, with ^{13}C isotropic shifts of $\delta = 250, 148, 48$, and 24. This species is seen to be an intermediate in the formation of another seven-carbon species, i.e. toluene.

Each NMR spectrum in Figs. 3.12 and 3.13 required the preparation of a new sample. In each case, this required two or three hours to prepare the reactor, a few seconds for the chemistry, an hour to transfer the catalyst bed to the NMR instrument, and then another hour of spectral acquisition to obtain very high signal-to-noise. A very experienced and efficient postdoctoral researcher would need three or four days to duplicate the eleven spectra in these two figures. In comparison, the variable-temperature CAVERN experiment in Fig. 3.5 required a single catalyst sample, and less than three hours from methanol adsorption to completion to measure twenty or more unique spectra (five are shown in the Fig.). Furthermore, in the CAVERN experiment, the spectra were acquired at reaction temperature while the reaction was underway, while in the pulse-quench experiments the catalysis and the spectroscopy took place in different rooms and sometimes on differ-

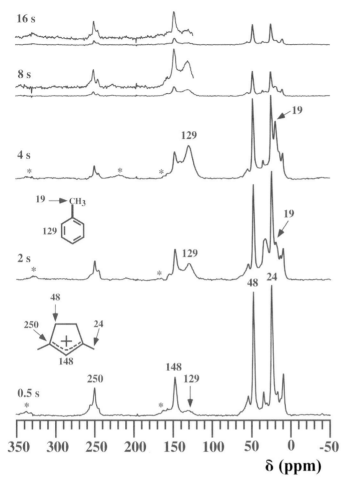

Fig. 3.12 ^{13}C CP/MAS NMR spectra of the reaction products of [^{13}C$_2$]ethylene retained on zeolite HZSM-5 following various times at 623 K in a pulse-quench reactor. Signals from cyclopentyl cations (δ = 250, 148, 48, and 24) and toluene (δ = 129 and 19) are indicated in the spectra. All spectra were measured at 298 K. The asterisks denote spinning sidebands.

ent days. The quench reactor approach to in situ NMR might therefore seem like a giant leap backwards; it fails entirely to meet the most restrictive definition of an in situ experiment, and it requires a lot more effort (and isotope) per distinct NMR spectrum than a conventional sealed rotor experiment. However, without quench experiments it is doubtful that in situ NMR could have contributed to the elucidation of the methanol conversion mechanism.

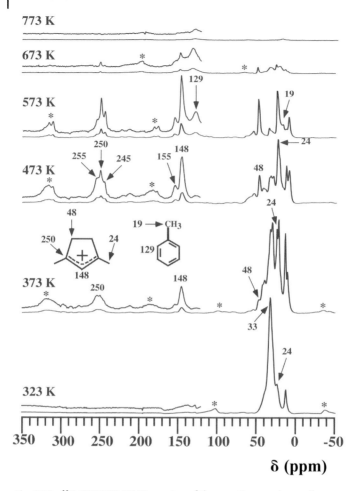

Fig. 3.13 ^{13}C CP/MAS NMR spectra of the reaction products of [^{13}C$_2$]ethylene retained on zeolite HZSM-5 following 4 s of reaction at various temperatures (indicated) in a pulse-quench reactor. Toluene ($\delta = 129$ and 19) is one of the products formed from the cyclopentenyl carbenium ion at higher temperatures. All spectra were measured at 298 K. The asterisks denote spinning sidebands.

3.3
Applications of In Situ NMR to Methanol-to-Olefin Catalysis

3.3.1
Overview

The conversion of methanol to olefins (MTO) and other hydrocarbons on microporous solid acids, especially the aluminosilicate HZSM-5 and more recently the

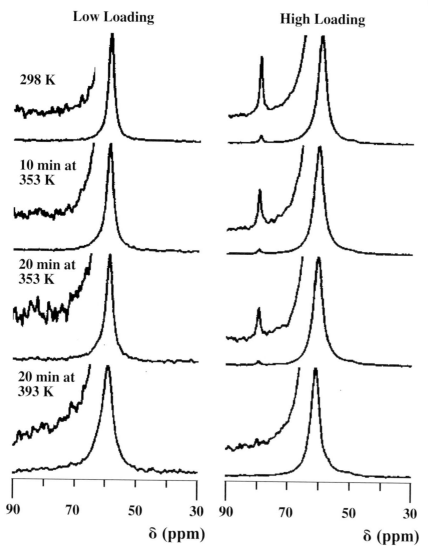

Fig. 3.14 ¹³C MAS NMR in situ studies of the reactions of high (1.8 equiv./acid site) and low (0.2 equiv./acid site) loadings of [¹³C]dimethyl ether on zeolite HZSM-5. Trimethyloxonium formed only in the high loading sample and disappeared as the sample was heated to 393 K.

silico-aluminophosphate HSAPO-34, is one of the most extensively studied problems in catalysis [66, 67]. A recent review cited 350 papers, a fraction of the total, and identified more than 20 distinct mechanistic proposals for the reaction by which methanol and/or dimethyl ether (DME) somehow form the "first" C–C bond [68]. Some of these mechanisms can be grouped into classes based on

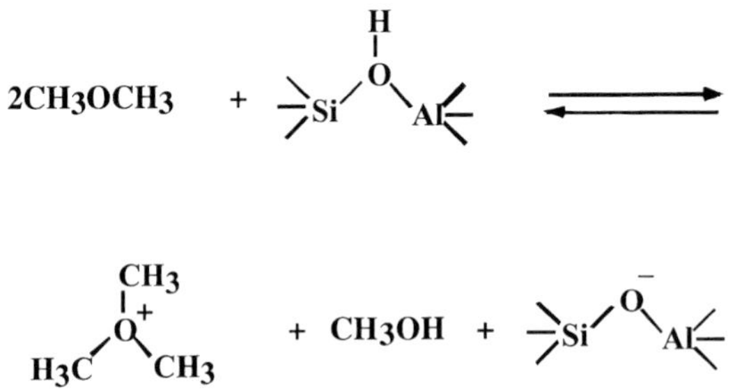

$$2CH_3OCH_3 \quad + \quad \underset{Si}{\overset{H}{\underset{}{\bigg\rangle}}}\overset{O}{\diagup}\overset{}{\diagdown}Al\bigg\langle \quad \xrightleftharpoons{\qquad}$$

$$\underset{H_3C}{\overset{CH_3}{\underset{}{\bigg|}}}\overset{+}{O}\diagdown CH_3 \quad + \quad CH_3OH \quad + \quad \underset{Si}{\bigg\rangle}\overset{\overset{-}{O}}{\diagup}\diagdown Al\bigg\langle$$

Scheme 3.1

whether the key intermediates are oxonium ylides, carbenes, carbocations, free radicals, or surface-bound alkoxy species. All of these direct routes for the conversion of methanol or dimethyl ether to hydrocarbons involve the participation of a small number of carbon atoms, two to four, and the formation of ethylene, propene, or an oxygenate that immediately decomposes to an olefin under the reaction conditions.

Many in situ NMR experiments have been directed toward methanol conversion. Surface-bound alkoxy species have been unambiguously identified in a number of NMR studies of alcohols and olefins on zeolites and related materials, and these are consistent with one of the classical reaction mechanisms. The CAVERN experiments in Fig. 3.14 show that at high concentration, dimethyl ether disproportionates at room temperature to form the trimethyloxonium ion, a species essential to the classical oxonium-ylide mechanism (Scheme 3.1). However, these same experiments show that the reaction is reversed at higher temperatures, well below the onset of hydrocarbon synthesis [69, 70].

A handful of papers considered indirect routes from methanol/DME to olefins. In 1983, Mole and co-workers reported that deliberately introduced toluene acted as a "co-catalyst" for methanol conversion on HZSM-5 [71,72]. They proposed a mechanism by which methyl substituents on benzene rings underwent side-chain alkylation followed by olefin elimination. Kolboe later observed that the activity of either zeolite HZSM-5 or the aluminosilicate HSAPO-34 in methanol conversion was dependent upon the catalysts' prior exposure to ethylene or ethanol, and that hydrocarbon products from a second pulse of methanol incorporated some carbon label from a previous pulse of ethanol. Kolboe et al. proposed that a "hydrocarbon pool" of unspecified structure formed in the catalysts and that this undergoes methylation and olefin elimination [73, 74]. The hydrocarbon pool mechanism has proven to explain many details of the MTO mechanism, and as shown below, in situ NMR has provided many details about the structure and function of the hydrocarbon pool.

3.3.2
HSAPO-34 Catalyst Structure

HSAPO-34 is a silico-aluminophosphate of the chabazite (CHA) structure. The CHA topology features cages that are interconnected through windows formed with eight tetrahedral (T = Al, P, or Si) sites. These are called eight-rings; in comparison, zeolite HZSM-5 (not shown) has larger channels composed of ten-rings.

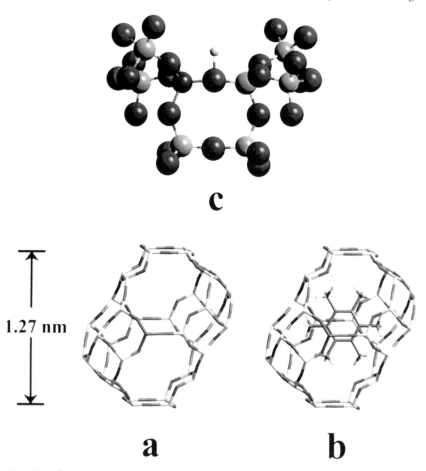

Fig. 3.15 Illustrations of the HSAPO-34 methanol-to-olefin catalyst, which figures heavily in the applications that follow: **a.** View of the nanometer-size cage without organic species present. The CHA structure of the catalyst is formed by connecting these cages at their eight-ring windows. Since a single cage is shown, a bond is removed from each T-site, and only the three bonds to oxygens within the single cage are shown. **b.** The same cage is shown with a hexamethylbenzene molecule included. **c.** Ball-and-stick model showing those atoms in the vicinity of an acid site. This is formed by substitution of a P atom by Si, forming an Si–OH–Al linkage, as in aluminosilicate zeolites.

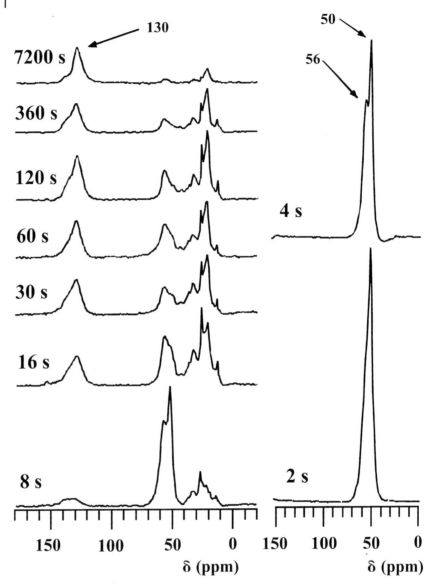

Fig. 3.16 75 MHz ^{13}C CP/MAS NMR spectra of samples from a pulse-quench study of methanol conversion on HSAPO-34 at 673 K. Each sample was prepared by injecting 20 µL of [^{13}C]methanol onto a freshly activated catalyst bed (0.3 g) while He was passed at 600 mL min^{-1}; reaction was allowed to proceed for the times shown and was terminated by a rapid thermal quench. All spectra (4000 scans) were measured at 298 K.

Reactants such as methanol and dimethyl ether and products such as ethylene and propene may freely diffuse through HSAPO-34, but products with larger kinetic diameters, even isobutylene, are trapped within the cages. Figure 3.15 shows the structure of this cage, as well as a similar cage containing a molecule of hexamethylbenzene (Fig. 3.15b). Clearly, aromatic molecules cannot pass through the eight-ring windows. A single acid site is shown in Figs. 3.15a and 3.15b; this is created by the replacement of a silicon atom by phosphorus. The choice of template molecule used to synthesize HSAPO-34 ensures that each cage has one and only one acid site. The local structure of the acid site is shown in the ball-and-stick model in Fig. 3.15c. This is the same Si-O-Al structure as for the acid site found in an aluminosilicate zeolite; in HSAPO-34 the T-sites neighboring the acid site are Al and P, whereas in an aluminosilicate zeolite they would all be silicon.

3.3.3
Pulse-Quench In Situ NMR of the Hydrocarbon Pool on HSAPO-34

Figure 3.16 shows ^{13}C MAS NMR spectra of HSAPO-34 samples that were exposed to identical pulses of [^{13}C]methanol for between 2 s and 7200 s in a flowing He stream at 673 K prior to a rapid thermal quench [75]. In this experiment, the product gases were also analyzed by GC. At the shortest reaction times, the conversion of methanol and dimethyl ether was low, and the ^{13}C NMR spectra showed both a signal due to adsorbed methanol ($\delta = 50$) and a shoulder at $\delta = 56$. The latter signal is attributable to a chemisorbed methoxy (methoxonium) species formed by replacing the proton at the acid site by a methyl group. After reaction for 4 s, a dramatic reduction in the amount of methanol on the catalyst was seen and methyl-substituted aromatics had formed, as indicated by the aromatic resonance at $\delta = 129$ (with a shoulder at $\delta = 134$ due to substituted ring carbons) and a methyl group signal at $\delta = 20$. Other upfield signals may be attributed to alkane products trapped in the cages. The average number of methyl groups per ring reached a maximum of ca. 4 between 30 and 120 s of reaction, and then decreased to ca. 1.4 after the catalyst aged for 7200 s at 673 K without injection of additional feed. GC analysis confirmed that traces of olefin products exited the catalyst bed even 7200 s after methanol injection.

Figure 3.17 shows GC traces that compare the activity of a completely fresh catalyst exposed to single pulse of methanol with that of a second catalyst that was first treated with a methanol pulse 360 s prior to a second, identical pulse. With the fresh catalyst, the conversion was only 14%, but the catalyst pretreated to form methylbenzenes achieved nearly 100% conversion of the second methanol pulse. A control experiment in Fig. 3.17 shows that the products from the second pulse vastly overwhelm those from the first pulse after 360 s of reaction. Experiments such as those relating to Figs. 3.16 and 3.17 provided very strong evidence that in

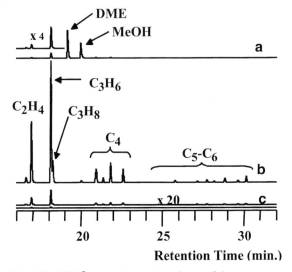

Fig. 3.17 GC (flame ionization) analyses of the gases exiting the HSAPO-34 catalyst bed operated as in the previous figure, except as a "double-pulse" experiment. Identical, 20 µL methanol pulses were applied at 0 s and at 360 s: **a.** 4 s after a first pulse the total conversion of methanol and dimethyl ether (DME) to hydrocarbons was only ca. 14%. **b.** 364 s after the first pulse and 4 s after the second pulse, the conversion was essentially 100%. **c.** This control experiment shows that only traces of products exit the reactor 358 s after the first methanol pulse; hence, the products observed at 364 s reflect conversion of the second methanol pulse.

order for HSAPO-34 to be active for MTO chemistry, it must have methylbenzenes trapped in some of its cages. A rapid succession of papers from several groups established that methylbenzenes are the hydrocarbon pool species in MTO catalysis on HSAPO-34.

3.3.4
Correlation of Product Selectivity (GC) with Catalyst Structure (NMR)

Figures 3.18 to 3.20 relate to a more detailed study of the methylbenzene species active in MTO catalysis on HSAPO-34 [76]. Using a motorized syringe pump, methanol was introduced at a weight hourly space velocity of 8 h^{-1} onto an HSAPO-34 catalyst bed at 673 K. After delivering 0.1 mL of methanol onto a 300 mg catalyst bed, the methanol flow was abruptly terminated and the gas stream exiting the reactor was repeatedly sampled for GC analysis. Figure 3.18

shows, not surprisingly, that the overall yield of olefins decreased significantly as the catalyst aged with no further reagent introduction. Figure 3.18 also shows that there is a significant change in product selectivity during this ageing process; immediately after termination of methanol flow, propene was the predominant product, but ethylene selectivity increased with ageing. Figure 3.19 shows "flow-wait-quench" NMR results from experiments designed to probe the evolution of catalyst structure during the ageing process resulting in Fig. 3.18. In each case, the catalyst bed reached a steady state with continuous methanol flow at 8 h^{-1}; the methanol flow was then terminated, the catalyst was maintained at reaction temperature with carrier gas flow for the time indicated, and then the temperature was quenched. The essential trend in Fig. 3.19 is that the average number of methyl groups per benzene ring decreased during catalyst ageing. Figure 3.20 combines the results of the two previous figures to correlate product selectivity (from GC analysis) with the average number of methyl groups per benzene ring (from quench NMR). This correlation shows that methylbenzenes with four or more methyl groups favor propene, while ethylene is favored with three or fewer methyl groups per ring. If it were to prove generally feasible to correlate in situ NMR results with product analysis, NMR could make a major impact on catalysis research.

The insight afforded by the results in Figs. 3.18 to 3.20 suggests ways of adjusting the process conditions to rationally alter MTO product selectivity. Recall that ethylene is favored by hydrocarbon pool methylbenzenes with fewer methyl

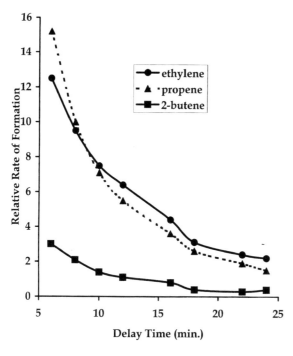

Fig. 3.18 The rates of formation of ethylene, propene, and 2-butene as a function of time from a single experiment similar to those used to generate Fig. 3.19. For these measurements, multiple gas samples were analyzed from a single catalyst bed as it evolved over time after cessation of methanol flow. Ethylene and propene are major primary products of MTO chemistry.

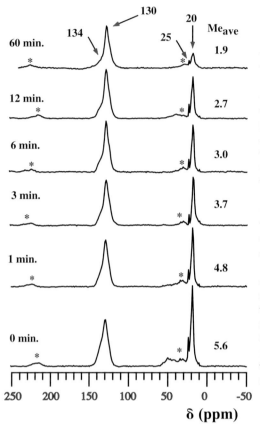

Fig. 3.19 75 MHz ^{13}C CP/MAS NMR spectra showing the loss of methyl groups as a function of time from methylbenzenes trapped in the HSAPO-34 nanocages at 400°C. For each spectrum, a fresh catalyst bed was used to convert 0.1 mL of [^{13}C]methanol at a WHSV of 8 h^{-1}, and then methanol flow was abruptly cut off. The catalyst bed was maintained at temperature with He flow (200 sccm) for the time indicated, and then the reactor temperature was rapidly quenched to ambient. Entire catalyst beds were loaded into MAS rotors to avoid sampling errors, and cross-polarization spectra were measured at room temperature. The average numbers of methyl groups per ring, Me$_{ave}$, were calculated from Bloch decay spectra very similar to the cross-polarization spectra shown.

groups per ring. Under steady-state conditions, the average number of methyl groups per ring reflects a balance between methylation and olefin elimination, and the rate of methylation might be reduced by lowering the methanol space velocity. This proved to be the feasible. Figure 3.21 shows a plot of ethylene selectivity against the logarithm of the space velocity. At very low space velocities, remarkably high ethylene selectivities were achieved. Industrial space velocities must be high for an economical process, so the results in Fig. 3.21 cannot be put into practice. Nevertheless, this is an example in which insights from in situ NMR experiments have allowed us to rationally alter catalytic process conditions (at the bench scale) to achieve a desirable end.

3.4
Some Limitations of In Situ NMR

The dream of the in situ spectroscopist is to conduct an uncompromised industrial process inside an all-seeing instrument with infinite sensitivity and resolving

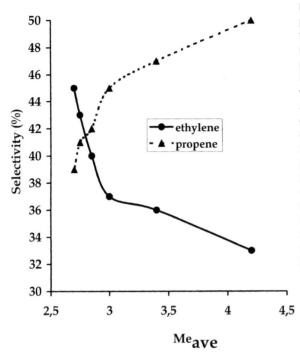

Fig. 3.20 Ethylene and propene selectivity as a function of the average number of methyl groups per ring, Me_{ave}. The data in Figs. 3.18 and 3.19 were plotted using an abscissa derived by empirically fitting the time evolution of Me_{ave} in Fig. 3.19 to a smooth curve generated by Excel. This mapping shows that ethylene is favored by methylbenzenes with two or three methyl groups, while propenes is favored with four or more methyl groups.

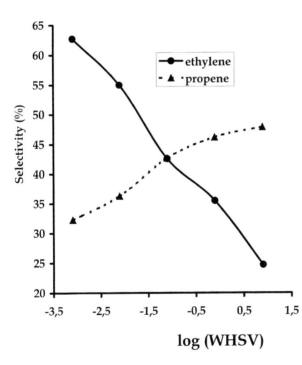

Fig. 3.21 Ethylene selectivity at 673 K *vs.* the logarithm of methanol weight-hourly space velocity. The catalyst bed, 300 mg, was first treated with a total of 0.1 mL of methanol at a WHSV of 8 h^{-1} to create methylbenzenes, and then was held at a given space velocity until selectivity reached a steady-state value. Note the very high ethylene selectivities at very low space velocities.

power. NMR is a great technique with a wonderful future in catalysis, but it cannot work alone. Some obvious limits of in situ NMR include ferromagnetic catalysts, and the study of paramagnetic metal oxides by NMR can be very challenging. Particles of non-ferrous metals can have large anisotropic bulk magnetic susceptibilities. Even so, the groups of Oldfield [77] and Reimer [78] have reported some very nice examples of in situ ^{13}C NMR of adsorbates on the metal particles of fuel cell catalysts.

Even with the best of samples, NMR is one of the least sensitive forms of spectroscopy. In situ NMR works because zeolites and many other catalytic materials

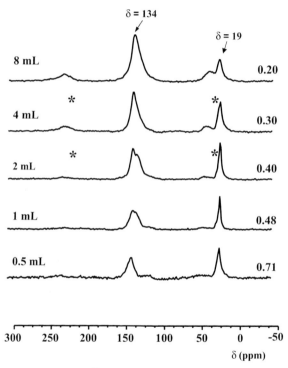

Fig. 3.22 75 MHz ^{13}C MAS NMR spectra (Bloch decays) of HSAPO-34 catalysts formed by converting various amounts of [^{13}C]methanol until deactivation eventually occurred. The aromatic hydrocarbons trapped in the HSAPO-34 cages give rise to the aromatic carbon signal at $\delta \approx 134$ and the methyl carbon signal at $\delta = 19$. While it is clear that the total content of aromatics increased with time on stream, the identities and relative compositions of various compounds cannot be deduced from these spectra; * denotes spinning sidebands. Numbers to the right of the spectra are the ratios of methyl carbons to aromatic ring carbons from the integrated intensities.

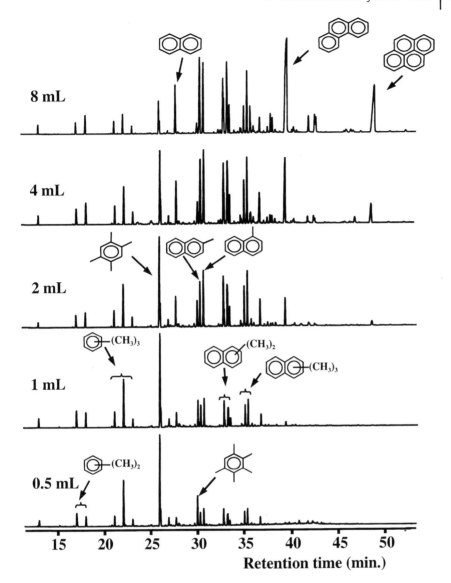

Fig. 3.23 GC-MS total ion chromatograms from analyses of CCl$_4$ extracts of cryogenically ground samples from HSAPO-34 catalyst beds treated with increasing amounts of methanol at 673 K. The NMR spectra of these samples are shown in Fig. 3.22. While methylbenzenes are the major species present early in the lifetime of the catalyst, methylnaphthalenes are also significant after one or two mL of methanol. Phenanthrene and pyrene are prominent on the deactivated catalyst.

have tremendous surface areas, hundreds of m^2 per gram of material. NMR is not usually applicable to the well-defined two-dimensional surfaces that are the subject of UHV surface science.

Those are the obvious limits of NMR. More generally, one has to decide whether the point is to learn about catalysis or to practice ones favorite spectroscopy. Figures 3.16 and 3.18 show ^{13}C NMR spectra of aromatic compounds, mostly methylbenzenes, that form in HSAPO-34 catalysts during methanol conversion. Other aromatic ring systems also form in HSAPO-34, especially as the catalyst ages and deactivates. Unfortunately, the ^{13}C NMR spectra of methylnaphthalenes are not obviously different from those of methylbenzenes, at least not with the resolution achievable in solids. The NMR spectra in Fig. 3.18 could be interpreted as measures of the average numbers of methyl groups per aromatic ring, but not as giving a more detailed distribution of methylbenzene isomers.

As the last result in this chapter, compare the in situ NMR study in Fig. 3.22 with the ex situ study of the same samples in Fig. 3.23. A series of HSAPO-34 catalysts were prepared by converting increasingly large amounts of methanol and then quenching the catalyst bed temperature. Here, larger methanol volumes were used in some experiments, and the catalyst was almost totally deactivated after converting the largest volume. The quench NMR spectra in Figure 3.22 do not reveal very much about the changing composition of the organic matter on the catalyst during time on stream leading to deactivation. A freezer mill was used to destroy the inorganic catalyst matrix in these samples and to liberate the organic material for ex situ analysis [79]. Freeze milling is used forensically to liberate DNA samples from bone and teeth. It is a very gentle method. CCl_4 extracts of these samples were analyzed by GC-MS, and Fig. 3.23 shows total ion chromatograms from these analyses. Far more information is provided by the ex situ analysis than by the in situ spectroscopic study. The point here is that we are all ultimately concerned with understanding catalysis and not with promoting one technique or another as a panacea.

Acknowledgements

The author thanks Profs. D. Raftery of Purdue University and E. Munson of the University of Kansas for permission to include graphics from their work in this chapter. Dr. W. Song and graduate student David Marcus of the University of Southern California provided invaluable assistance in assembling the other graphics. Financial support for this work was provided by the National Science Foundation and the Department of Energy.

References

1 Fyfe, C. A.; Feng, Y.; Grondey, H.; Kokotailo, G. T.; Gies, H., *Chem. Rev. 1991,* **91**, 1525–1543.

2 Fyfe, C. A.; Grondey, H.; Feng, Y.; Kokotailo, G. T., *J. Am. Chem. Soc. 1990,* **112**, 8812–8820.

3 Fyfe, C. A.; Bretherton, J. L.; Lam, L. Y., *J. Am. Chem. Soc. 2001,* **123**, 5285–5291.

4 Fyfe, C. A.; Bretherton, J. L.; Lam, L. Y., *Chem. Commun. 2000,* 1575–1576.

5 Beck, L. W.; White, J. L.; Haw, J. F., *J. Am. Chem. Soc. 1994,* **116**, 9657–9661.

6 Liu, H. M.; Kao, H. M.; Grey, C. P., *J. Phys. Chem. B 1999,* **103**, 4786–4796.

7 White, J. L.; Beck, L. W.; Haw, J. F., *J. Am. Chem. Soc. 1992,* **114**, 6182–6189.

8 Xu, T.; Munson, E. J.; Haw, J. F., *J. Am. Chem. Soc. 1994,* **116**, 1962–1972.

9 Haw, J. F.; Chuang, I.-S.; Hawkins, B. L.; Maciel, G. E., *J. Am. Chem. Soc. 1983,* **105**, 7206–7207.

10 Baltusis, L.; Frye, J. S.; Maciel, G. E., *J. Am. Chem. Soc. 1986,* **108**, 7119–7120.

11 Lunsford, J. H.; Tutunjian, P. N.; Chu, P. J.; Yeh, E. B.; Zalewski, D. J., *J. Phys. Chem. 1989,* **93**, 2590–2595.

12 Kao, H. M.; Liu, H. M.; Jiang, J. C.; Lin, S. H.; Grey, C. P., *J. Phys. Chem. B 2000,* **104**, 4923–4933.

13 Kao, H. M.; Grey, C. P., *J. Am. Chem. Soc. 1997,* **119**, 627–628.

14 Haw, J. F.; Zhang, J. H.; Shimizu, K.; Venkatraman, T. N.; Luigi, D. P.; Song, W. G.; Barich, D. H.; Nicholas, J. B., *J. Am. Chem. Soc. 2000,* **122**, 12561–12570.

15 Bonardet, J. L.; Fraissard, J.; Gedeon, A.; Springuel-Huet, M. A., *Catal. Rev. Sci. Eng. 1999,* **41**, 115–225.

16 Medek, A.; Harwood, J. S.; Frydman, L., *J. Am. Chem. Soc. 1995,* **117**, 12779–12787.

17 Marinelli, L.; Medek, A.; Frydman, L., *J. Magn. Reson. 1998,* **132**, 88–95.

18 Bodart, P. R.; Amoureux, J. P.; Pruski, M.; Bailly, A.; Fernandez, C., *Magn. Reson. Chem. 1999,* **37**, S69–S74.

19 Lim, K. H.; Grey, C. P., *J. Am. Chem. Soc. 2000,* **122**, 9768–9780.

20 Bull, L. M.; Cheetham, A. K.; Anupold, T.; Reinhold, A.; Samoson, A.; Sauer, J.; Bussemer, B.; Lee, Y.; Gann, S.; Shore, J.; Pines, A.; Dupree, R., *J. Am. Chem. Soc. 1998,* **120**, 3510–3511.

21 Haw, J. F.; Nicholas, J. B.; Xu, T.; Beck, L. W.; Ferguson, D. B., *Acc. Chem. Res. 1996,* **29**, 259–267.

22 Hunger, M.; Weitkamp, J., *Angew. Chem. Int. Ed. 2001,* **40**, 2954–2971.

23 Haw, J. F.; Xu, T., *Adv. Catal. 1998,* **42**, 115–180.

24 Haw, J. F., *Topics Catal. 1999,* **8**, 81–86.

25 Xu, T.; Haw, J. F., *Topics Catal. 1997,* **4**, 109–118.

26 Parker, W. O., *Comments Inorg. Chem. 2000,* **22**, 31–73.

27 Hwang, S. J.; Petucci, C.; Raftery, D., *J. Am. Chem. Soc. 1997,* **119**, 7877–7878.

28 Hwang, S. J.; Petucci, C.; Raftery, D., *J. Am. Chem. Soc. 1998,* **120**, 4388–4397.

29 Hwang, S. J.; Raftery, D., *Catal. Today 1999,* **49**, 353–361.

30 Pilkenton, S.; Hwang, S. J.; Raftery, D., *J. Phys. Chem. B 1999,* **103**, 11152–11160.

31 Xu, T.; Barich, D. H.; Goguen, P. W.; Song, W. G.; Wang, Z. K.; Nicholas, J. B.; Haw, J. F., *J. Am. Chem. Soc. 1998,* **120**, 4025–4026.

32 Xu, T.; Haw, J. F., *J. Am. Chem. Soc. 1994,* **116**, 7753–7759.

33 Xu, T.; Haw, J. F., *J. Am. Chem. Soc.* **1994**, *116*, 10188–10195.

33 Stepanov, A. G.; Sidelnikov, V. N.; Zamaraev, K. I., *Chem. Eur. J.* **1996**, *2*, 157–167.

34 Song, W. G.; Nicholas, J. B.; Haw, J. F., *J. Am. Chem. Soc.* **2001**, *123*, 121–129.

35 Beck, L. W.; Xu, T.; Nicholas, J. B.; Haw, J. F., *J. Am. Chem. Soc.* **1995**, *117*, 11594–11595.

36 Anderson, M. W.; Klinowski, J., *Nature* **1989**, *339*, 200–203.

37 Anderson, M. W.; Klinowski, J., *J. Am. Chem. Soc.* **1990**, *112*, 10–16.

38 Anderson, M. W.; Sulikowski, B.; Barrie, P. J.; Klinowski, J., *J. Phys. Chem.* **1990**, *94*, 2730–2734.

39 Derouane, E. G.; Hamid, S. B. A.; Pasauclaerbout, A.; Seivert, M.; Ivanova, I. I., *Sci. Techn. Catal.* **1994–1995**, *92*, 123–130.

40 Ivanova, I. I.; Blom, N.; Derouane, E. G., *Catal. Microporous Mater.* **1995**, *94*, 419–426.

41 Ivanova, I. I.; Rebrov, A. I.; Pomakhina, E. B.; Derouane, E. G., *J. Molec. Catal. A: Chemical* **1999**, *141*, 107–116.

42 Munson, E. J.; Ferguson, D. B.; Kheir, A. A.; Haw, J. F., *J. Catal.* **1992**, *136*, 504–509.

43 Munson, E. J.; Murray, D. K.; Haw, J. F., *J. Catal.* **1993**, *141*, 733–736.

44 Haw, J. F.; Richardson, B. R.; Oshiro, I. S.; Lazo, N. D.; Speed, J. A., *J. Am. Chem. Soc.* **1989**, *111*, 2052–2058.

45 Munson, E. J.; Kheir, A. A.; Lazo, N. D.; Haw, J. F., *J. Phys. Chem.* **1992**, *96*, 7740–7746.

46 Oliver, F. G.; Munson, E. J.; Haw, J. F., *J. Phys. Chem.* **1992**, *96*, 8106–8111.

47 Ernst, H.; Freude, D.; Mildner, T., *Chem. Phys. Lett.* **1994**, *229*, 291–296.

48 Mildner, T.; Ernst, H.; Freude, D.; Holderich, W. F., *J. Am. Chem. Soc.* **1997**, *119*, 4258–4262.

49 Ernst, H.; Freude, D.; Mildner, T.; Wolf, I., *Solid-State Nucl. Magn. Reson.* **1996**, *6*, 147–156.

50 Ferguson, D. B.; Haw, J. F., *Anal. Chem.* **1995**, *67*, 3342–3348.

51 Haddix, G. W.; Reimer, J. A.; Bell, A. T., *J. Catal.* **1987**, *106*, 111–115.

52 Hunger, M.; Horvath, T., *J. Chem. Soc., Chem. Commun.* **1995**, 1423–1424.

53 Hunger, M.; Horvath, T., *J. Catal.* **1997**, *167*, 187–197.

54 Hunger, M.; Horvath, T.; Weitkamp, J., *Microporous Mesoporous Mater.* **1998**, *22*, 357–367.

55 Wang, W.; Seiler, M.; Ivanova, I. I.; Weitkamp, J.; Hunger, M., *Chem. Commun.* **2001**, 1362–1363.

56 Isbester, P. K.; Kaune, L.; Munson, E. J., *Chemtech* **1999**, *29*, 40–47.

57 Isbester, P. K.; Zalusky, A.; Lewis, D. H.; Douskey, M. C.; Pomije, M. J.; Mann, K. R.; Munson, E. J., *Catal. Today* **1999**, *49*, 363–375.

58 Goguen, P.; Haw, J. F., *J. Catal.* **1996**, *161*, 870–872.

59 Haake, M.; Pines, A.; Reimer, J. A.; Seydoux, R., *J. Am. Chem. Soc.* **1997**, *119*, 11711–11712.

60 Keeler, C.; Xiong, J. C.; Lock, H.; Dec, S.; Tao, T.; Maciel, G. E., *Catal. Today* **1999**, *49*, 377–383.

61 Macnamara, E.; Raftery, D., *J. Catal.* **1998**, *175*, 135–137.

62 Bax, A.; Szeverenyi, N. M.; Maciel, G. E., *J. Magn. Reson.* **1983**, *52*, 147–152.

63 Haw, J. F.; Goguen, P. W.; Xu, T.; Skloss, T. W.; Song, W. G.; Wang, Z. K., *Angew. Chem. Int. Ed.* **1998**, *37*, 948–949.

64 Goguen, P. W.; Xu, T.; Barich, D. H.; Skloss, T. W.; Song, W. G.; Wang, Z. K.; Nicholas, J. B.; Haw, J. F., *J. Am. Chem. Soc.* **1998**, *120*, 2650–2651.

65 Haw, J. F.; Nicholas, J. B.; Song, W. G.; Deng, F.; Wang, Z. K.; Xu, T.; Heneghan, C. S., *J. Am. Chem. Soc.* **2000**, *122*, 4763–4775.

66 Chang, C. D., *Catal. Rev. 1983*, **25**, 1–118.

67 Keil, F. J., *Microporous Mesoporous Mater. 1999*, **29**, 49–66.

68 Stöcker, M., *Microporous Mesoporous Mater. 1999*, **29**, 3–48.

69 Munson, E. J.; Haw, J. F., *J. Am. Chem. Soc. 1991*, **113**, 6303–6305.

70 Munson, E. J.; Kheir, A. A.; Haw, J. F., *J. Phys. Chem. 1993*, **97**, 7321–7327.

71 Mole, T.; Whiteside, J. A.; Seddon, D., *J. Catal. 1983*, **82**, 261–266

72 Mole, T.; Bett, G.; Seddon, D., *J. Catal. 1983*, **84**, 435–445.

73 Dahl, I. M.; Kolboe, S., *Catal. Lett. 1993*, **20**, 329–336.

74 Dahl, I. M.; Kolboe, S., *J. Catal. 1994*, **149**, 458–464.

75 Song, W. G.; Haw, J. F.; Nicholas, J. B.; Heneghan, C. S., *J. Am. Chem. Soc. 2000*, **122**, 10726–10727.

76 Song, W. G.; Fu, H.; Haw, J. F., *J. Am. Chem. Soc. 2001*, **123**, 4749–4754.

77 Tong, Y. Y.; Rice, C.; Wieckowski, A.; Oldfield, E., *J. Am. Chem. Soc. 2000*, **122**, 1123–1129.

78 Yahnke, M. S.; Rush, B. M.; Reimer, J. A.; Cairns, E. J., *J. Am. Chem. Soc. 1996*, **118**, 12250–12251.

79 Fu, H.; Song, W.; Haw, J. F., *Catal. Lett., 2001*, **76**, 89–94.

4
Theoretical Catalysis: Methods, Applications, and Future Directions

John B. Nicholas

4.1
Introduction

While there has been great progress in the development of in situ experimental methods for the study of catalysis, there have also been great improvements in the theoretical study of catalysis. The earliest studies of catalytic systems were severely limited by the slow computers of the time, and generally inefficient algorithms. However, as is well known, computer speed has increased by orders of magnitude, while the cost of computers has decreased by similar amounts. The speed of the now very common thousand dollar desktop computer is in many cases greater than that of the computers that used to run whole departments or universities, costing many hundreds of thousands or millions of dollars. Along with these huge increases in computer power, there have also been great improvements in the algorithms used in theoretical chemistry. Finally, the development of vector and parallel computers, and the techniques needed to exploit them, have increased effective performance by several more orders of magnitude. The result of these developments is that we can now theoretically model systems large enough to be relevant to catalysis, with an accuracy that can provide quantitative predictions of many chemical properties.

The relationship between theoretical methods and in situ experimental methods for the study of catalysis is complex. The obvious goal of in situ experiments is to study catalysis under reaction conditions. In some cases, theoretical methods can simulate behavior under "reaction conditions", typically by incorporating realistic temperature and pressure effects. Of course, the major difference between a theoretical and experimental measurement is the number of atoms involved. Theoretical methods simply cannot model the 10^{23} atoms studied by typical experimental techniques. Thus, the theoretical chemist must resort to some simpler model of the true experimental system. Fortunately, the accuracy of a theoretically calculated IR, Raman, or NMR spectrum, based on even a simple model system, is usually more than adequate to prove or disprove experimental assignments. Perhaps the best way to view the relationship is to consider theoretical chemistry as a valuable complement to any experimental study, rather than as specifically an ex situ or in situ technique.

This chapter first presents the methods used in the theoretical study of catalysis. This discussion is broad, hopefully giving the reader an idea of what theoretical techniques are available and applicable to the study of catalysis. Included is an outline of the kinds of chemical properties that can be calculated, and the factors that influence accuracy. As the primary audience for this book is the experimental community, the discussion of theoretical methods does not cover their mathematical foundation. Finally, a series of examples of the use of theoretical methods in catalysis is presented. Much of this theoretical work has been done in collaboration with solid-state MAS NMR experiments conducted by Prof. James F. Haw and his group at the University of Southern California. The accuracy and utility of theoretical methods is proven by the fact that theoretical and experimental techniques have been able to complement each other on a wide range of problems [1].

4.2
Theoretical Methods

For the study of catalysis, theoretical methods can be broadly divided into those based on classical mechanics and those based on quantum mechanics. Although this chapter focuses largely on quantum mechanics, classical mechanics has had a large impact on catalysis, as is also discussed. All the methods discussed in this chapter work with an atomic level description (coordinates) of the system of interest. Thus, they are distinct from other types of computational models that deal with catalysis on a more macroscopic level.

4.2.1
Classical Mechanics

4.2.1.1 Force fields

All classical mechanical methods depend on a force field to represent interatomic and intermolecular interactions [2], such as bond stretches and bends, rotation about dihedral angles, and electrostatic and dispersive forces. The force field consists of a set of mathematical functions (such as harmonic potentials) and related force constants. As the values used in force fields do not generally correspond to true force constants, they are often referred to as force field parameters. Obviously, the accuracy of the results obtained with force fields depends largely on the accuracy with which the force field can reproduce chemical interactions. Thus, the parameterization of force fields is an important aspect of their use [3–5].

The interactions between molecules can be represented by Lennard-Jones potentials, which include short range, repulsive forces (related to the Pauli exclusion principle) and longer range, attractive forces arising from interactions between induced dipoles. Lennard-Jones potentials were first used to model rare gases. The parameterization of Lennard-Jones interactions has expanded over the years to include other gases, as well as liquids, polymers, and proteins [6–8], and catalytic systems such as clays [9, 10], zeolites [11], and other metal oxides [12]. Perhaps the earliest use of force field methods in catalysis appeared in Kiselev's attempts at modeling heats of adsorption and Henry's Law constants for zeolite-adsorbate systems [13]. As force fields became more widely used in catalysis, other parameterizations emerged, particularly for zeolites [11, 14, 15].

Obviously, there are other important chemical interactions that are not modeled well by simple Lennard-Jones potentials. In particular, in almost all cases, very long range electrostatic interactions must also be included [16, 17]. These interactions can be modeled in terms of Coulomb's law and partial charges located at the atomic centers. As the partial charge on an atom is not rigorously defined quantum mechanically, partial charges are often simply treated as adjustable parameters, refined to reproduce interatomic interaction energies. As the electrostatic potential is well defined, partial charges can also be obtained by fitting their values, through least-squares procedures, to reproduce the electrostatic potential of a model system [18]. The electrostatic potential is obtained by quantum mechanical calculations. Finally, higher order multipole moments (dipoles, quadrupoles, etc.) may be included in the force field to more accurately model electrostatic interactions. As electrostatic interactions are operative over a very long range, obtaining convergence requires special techniques, such as the Ewald summation [19] and the more efficient particle-mesh Ewald method [20, 21].

Molecular flexibility can also be modeled by force field methods. In this case, one would need parameters to represent the energies of bond stretches, angle bends, rotation about dihedral angles, and perhaps the coupling between these internal degrees of freedom. Such valence force field parameters for organic systems are widely available [2, 6–8]. Force fields designed to model the internal flexibility of inorganic catalysts, such as zeolites, are also available [11, 22, 23]. Valence force field parameters can be obtained by fitting to IR and Raman spectra, from empirical relationships (i.e. Badger's rule), and from quantum mechanical calculations of model systems. Many force fields do not explicitly include polarization, although this may be included implicitly by the adjustment of force field parameters. For example, accurate simulations of water generally use force fields for which the calculated gas-phase dipole moment is parameterized to be close to the experimentally measured dipole in solution. However, explicit polarization is increasingly included in force fields. This can be achieved by including point dipoles [24–27], partial charges that depend on molecular geometry (fluctuating charge models) [28], and the shell model [29].

4.2.1.2 Classical mechanical techniques

Energy minimization

As stated above, all force fields include potential functions that represent the energies of interatomic and intermolecular interactions. These functions are generally continuous (at least for the range of interactions that we are interested in) and differentiable. By taking the derivative of the energy with respect to the internal or Cartesian coordinates, the forces can be obtained. Analytical second derivatives of most force fields are also available. Given energies and derivatives, many standard optimization procedures can be used to obtain the minimum-energy geometries of individual molecules or molecular systems. The use of force fields and energy minimization techniques, with which the term molecular mechanics is often associated, was widely popularized by Allinger [2]. In cases in which initial geometries are crude and the forces are large, the steepest descents method, which relies only on energies and first derivatives, will reliably and quickly approach the minimum. However, steepest descents becomes much less efficient close to the minimum, where the potential energy surface is flatter but forces are still large, and it is not generally used to obtain the final geometry. A reliable method for refining geometries obtained by steepest descents is the conjugate gradient method, which uses information from prior search steps to guide the minimization path. Methods that utilize the full second derivative matrix (Hessian) are even more efficient at locating the energy minimum. These Newton-Rhapson methods converge very rapidly if the geometry is in a region of the potential energy surface that is quadratic. However, inversion of the Hessian matrix can be costly for large systems and the memory needed can also become quite large. A variety of other methods that approximate the Hessian are also available [30].

Starting from an initial position on the potential energy surface, energy minimization techniques move downhill to the nearest local minimum. In many cases, a variety of minimum-energy geometries exist, and a search for the global minimum may be necessary. The location of global minima is a classic problem in optimization, and still remains an area of active research [31]. Several techniques are commonly used in attempts to locate global minima, or at least to sample many local minima. Systematic or random sampling offers the potential to locate global minima, although both can be very costly in terms of computer time, and there is no assurance that the global minimum will be found (or recognized as such if it is found). More sophisticated techniques for locating global minima continue to be developed. Of these, Monte Carlo and molecular dynamics techniques are discussed below. Classical mechanical energy minimization can be used to optimize structures of catalysts, such as crystalline zeolites [32], and to locate preferred sites for adsorbed molecules within the zeolite channels.

Monte Carlo simulation

Monte Carlo (MC) methods can be used to explore conformational or configurational space. Current MC methods follow from the first computer simulations performed in 1953 [33]. In the MC technique, the system under study is subjected to a series of random moves, which may include translation or rotation of molecules, rotation about torsions, or even the addition or deletion of molecules (Grand Canonical Monte Carlo, GCMC). Starting from the initial configuration a move is made. If the perturbation of the system results in a lowering of the energy, the move is accepted. If the perturbation results in an increase in energy, the move is conditionally accepted. The quantity $\exp(-\Delta E/RT)$ is computed and compared to a random number between 0 and 1. If $\exp(-\Delta E/RT)$ is higher than the random number the move is accepted, otherwise it is rejected. By repeated application of this procedure, a sampling of a statistical mechanical ensemble is obtained. Typical MC methods simulate the canonical (constant N, V, and T) ensemble. Care must be taken that the proposed moves are not so small as to disallow the possibility of exploring configurational space, or so large that the moves are only rarely accepted, thus greatly slowing the sampling.

In order to model gases, liquids, or solids in a proper statistical mechanical ensemble, periodic boundary conditions are often used [34]. In simplest terms, a simulation box is defined, and molecules that effectively migrate out of one side of the box enter from the opposite side, allowing the number of molecules to remain constant, and hence defining a simulation volume and density. The evaluation of the force field terms is also carried out in a fashion that minimizes "edge effects" and allows each molecule to "feel" as if it is equally surrounded by the other molecules in the system. Periodic boundary conditions are also often used in molecular dynamics simulations (see below).

MC methods can be used to determine the distribution of adsorbed molecules within a zeolite [35–38], or to determine the preferred sites and distribution of aluminum within a silico-alumino framework. GCMC, which simulates a system at constant chemical potential, can be used to predict adsorption isotherms [39]. As alluded to above, MC methods can be used in attempts to locate global energy minima as they allow for the crossing of energy barriers.

Molecular dynamics simulation

Molecular dynamics (MD) methods simulate the time evolution of a system by analytically integrating Newton's equations of motion. The earliest MD simulations were those of Alder and hard spheres), followed by the simulation of Lennard-Jones particles by Rahman in 1964 [41]. Techniques are available by which the temperature [42] and pressure [43] may be controlled, thus providing access to several statistical-mechanical ensembles. The accuracy of the integration depends partially on the length of the MD time step, which, in turn, is limited by the speed of the fastest moving degrees of freedom within the system. As the fastest moving degrees of freedom are usually bond stretches involving hydrogen, the freezing of all C–H, O–H, N–H, etc. bonds is often accomplished using the efficient SHAKE procedure [44]. Freezing bonds involving hydrogen allows a time step of 1 femtosecond or larger. Current computers allow the routine simulation of modestly sized systems (~10,000 atoms) for nanoseconds. Simulations in the 200-nanosecond range have recently been reported. Simplifications of the potential energy function, such as treating CH_3 groups as single atoms (united atoms) or pre-calculating interaction energies for later evaluation by interpolation [45], can be used to decrease the computer time needed. As in the case of MC, periodic boundary conditions are often used to maintain a particular statistical mechanical ensemble.

MD simulations can be used in attempts to find a global minimum by simulated annealing [46]. In simulated annealing, an MD run is begun at high temperature, which should allow the system to surmount energy barriers with heights of approximately kT. The system is gradually cooled, in the hope that it will settle into the global energy minimum. While this method should find the global minimum in the limit of an infinitely slow cooling, the results for simulated annealing runs of practical length are less certain.

MD simulations have been used to study the time behavior of adsorbates on surfaces [12] and within channels of catalytic systems [47–52], and to study the internal flexibility (breathing motions) of zeolites and clays [53–56]. MD methods based on quantum mechanical energies and forces, rather than those from a classical force field, are now in common use (see below).

All of the classical mechanical techniques discussed above have been adapted to run efficiently on vector and parallel computers. Vector computers have special hardware registers that allow many computations to be performed at the same time, whereas parallel computers attain faster speed by coupling many processors together with some sort of fast network. Highly effective parallelization of MC simulations can be obtained by running independent simulations on different processors, and collating the results afterwards. A similar strategy for MD simulations is not generally available, as a single continuous run, corresponding to the desired amount of simulated time, is needed. In this case, the most CPU-intensive aspects of the simulation, calculation of the energies and forces, are divided between the processors [57]. Various parallel strategies have been devised, such as replicated data, spatial decomposition [58, 59], and force decomposition [60, 61]. Although effective parallelization of MD simulations is difficult to achieve, parallel computer codes are now available from both academic and commercial sources [58–62].

4.2.1.3 Prediction of experimental data with classical mechanical methods

The range of chemical properties that can be studied with classical methods is quite large. Classical methods can provide predictions of the structures and energetics of molecules of all sizes, from the small adsorbate molecules subject to reaction, to large inorganic catalysts such as zeolites, clays, and metal oxides. Predictions of vibrational frequencies can be made by energy minimization, followed by calculation, mass weighing, and diagonalization of the force constant matrix. From the force constant matrix, other properties, such as entropies and free energies, may also be estimated. The adsorption and diffusion of adsorbate molecules on surfaces and within the channels of porous solids can be simulated, providing orientations and distributions of molecules, heats of adsorption, and adsorption isotherms. Many properties can be obtained from MD with the use of correlation functions. Diffusion constants can also be obtained by Fourier transform of the velocity autocorrelation function (VACF), with the possibility of isolating translational or rotational diffusion constants. Fourier transform of the particle density correlation function provides predictions of scattering functions, whereas Fourier transform of the dipole correlation function provides the IR spectrum. Both MC and MD methods are often used to determine radial distribution functions. Readers interested in more details of the classical mechanical methods should consult the many excellent texts [63–65].

4.2.2
Quantum Mechanics

While classical methods depend on a force field to represent chemical interactions, quantum mechanical (QM) methods are based on first principles. Thus, many QM methods attempt a solution of the Schrödinger equation, and thus the determination of the wavefunction. The mathematics involved in this solution is complex and beyond the scope of this presentation. Interested readers can refer to many high quality texts [66–71]. Although QM methods can, in principle, calculate all molecular properties to very high accuracy, in practice the computer time needed for such calculations is far beyond current availability for most systems of interest to catalysis. Thus, the developers of QM techniques have invested much effort in simplifying approximations, writing more efficient algorithms, and allowing for adaptations to parallel computers.

4.2.2.1 Semiempirical methods

Semiempirical methods, as the name implies, depend partially on parameters adjusted to fit experimental data. It is worth noting that the number of classical force field parameters needed to study a complex system can be quite large (hundreds or thousands), as each interaction typically requires two or more parameters. Of the quantum mechanical methods, only semiempirical methods truly depend on a parameterization. However, in this case, there are only about ten parameters per element.

One of the most time-consuming parts of an electronic structure calculation is the evaluation of two-electron (overlap) integrals. Semiempirical methods attempts to save computing time by various approximations of the overlap integrals, the most drastic being the Complete Neglect of Differential Overlap (CNDO) method of Segal and Pople [72]. The INDO (Intermediate Neglect of Differential Overlap) and NDDO (Neglect of Diatomic Differential Overlap) methods were also developed [73, 74]. Another family of semiempirical methods was developed by Dewar and co-workers, who gave us MINDO/3 and MNDO [75, 76]. The most widely used semiempirical methods are currently AM1 (Austin Model 1) [77] developed by Dewar and co-workers, and PM3 [78, 79] developed by Stewart. Also, for the prediction of spectroscopic properties, the ZINDO method of Zerner and co-workers [80, 81] is popular.

Semiempirical methods can generally give accurate predictions of the properties of organic molecules, for which there is much experimental data to aid the parameterization. The accuracy provided for inorganic systems is lower, partially due to the lack of data for parameterization, and partially due to the limitations of the method. However, attempts are being made to more accurately parameterize and treat inorganic systems. The addition of d-orbitals (MNDO/d) [82] by Thiel and co-workers is a good step in this direction. Semiempirical methods have been used to study a variety of problems in catalysis, although more accurate ab initio and density functional methods (see below) are now much more widely used.

4.2.2.2 Ab initio methods

In contrast to semiempirical methods, ab initio methods do not depend on parameterization [69]. Thus, ab initio methods can be generally applied to any chemical system. Ab initio methods are also capable of greater accuracy than can typically be achieved with semiempirical techniques. The accuracy of ab initio methods depends greatly on the basis set used to define the atomic and molecular orbitals of the system, and the extent to which electron correlation is described. Hartree-Fock (HF) methods use an iterative, self-consistent field (SCF) approach to the solution of the equations. Both open and closed-shell systems can be treated. A major weakness of the HF approach is that it neglects electron correlation.

Within the ab initio framework, electron correlation is often included through Møller-Plessett perturbation theory, generally carried out to second (MP2), third (MP3), or fourth (MP4) order [83, 84]. Excitation of electrons from an occupied to an unoccupied (virtual) orbital are calculated with single, double, triple, and quadrupole excitations commonly available. Evidence indicates the triple excitations are often key to accuracy; the MP4(sdtq) method, which includes the triples, often provides high accuracy. Other methods for including higher-order correlation, such as coupled-cluster with single, double, and perturbative triple excitations (CCSD, CCSD(T)) are also available, and are widely regarded as being highly accurate [85]. Full configuration interaction (CI) approaches are also used [86]. It is important to note that the computational cost of these methods increases greatly as electron correlation is included to a greater extent. The computational cost is most closely related to the dependence of the computer time on the number of basis functions. All algorithms can be classified by both a formal scaling with basis set size (n), and a practical scaling, due to improvements in algorithm design. Thus, the RHF method has a formal scaling or n^4, but in practice performs as n^3 or better. The MPn and CC methods have a much larger n dependence. Thus, while it is possible to routinely treat 100's or 1000's of atoms at the RHF level on a large parallel computer, 100 atoms becomes a considerable problem at the MP2 or MP4 level, and 10's of atoms with CC methods.

As noted in the above discussion, the cost of the methods depends greatly on the basis set size, thus the basis set requirements should also be discussed. The basis set is a set of mathematical functions that represents the atomic orbitals. Basis sets commonly consist of Gaussian functions, although Slater and other functions are used. Each Gaussian function is created from a radial and an angular component, thus the functions correspond to our common concept of s, p, d, and higher angular momentum orbitals. A minimal basis set contains only enough functions to represent the filled valence electronic configuration of each atom. Thus, for carbon, the basis set would contain a 1s, 2s, and 2p function. The widely used STO-3G basis set is a minimal basis set. STO-3G signifies that three Gaussian functions, with different exponents and coefficients, are used to approximate a Slater orbital. For a given molecule, a minimal basis set affords the least number of functions, and with that, generally the least accuracy in the representation of atomic and molecular orbitals, and thus the least accuracy in terms of molecular properties. A much more reasonable basis set includes two sets of functions for each orbital, and is termed a double-zeta (DZ) basis set. Even more accuracy is afforded by triple-zeta basis sets, which obviously include three sets of functions per orbital. Even more complete basis sets, such as quadruple- and quintuple-zeta, are also used.

In many cases, polarization functions are added to the basis set. These are functions of higher angular moment than the valence configuration of the atom. Thus, for carbon, d-functions or higher serve as polarization functions. Polarization functions account for the deformation of atomic orbitals as molecular orbitals are formed. Anions, and molecules in which lone pairs or multiple bonds are important, often require the addition of diffuse functions, which provide the electrons with a more realistic spatial range. For many applications, a polarized double-zeta

(DZP) basis set is adequate. For highly accurate results, which by necessity require a method that includes electron correlation, a triple-zeta basis set with multiple polarization and diffuse functions may be needed. In practitioners terms, the double-zeta 6-31G* basis set of Pople and co-workers is often quite adequate [87–89], while the much larger aug-cc-pVTZ basis of Dunning and co-workers is often used for high accuracy [90, 91]. It should be noted that while large basis sets and higher-order correlation both provide paths to increased accuracy, it does not make sense to pursue one path alone. Thus, an RHF calculation with a very large basis set is unlikely to provide high accuracy as electron correlation is neglected. Similarly, a highly correlated calculation (i.e. CCSD(T)) with a minimal basis set might also give disappointing results as the basis set does not provide the flexibility needed to adequately represent electron correlation effects. If we now consider both the scaling of the QM methods and the basis set requirements, we can get a better idea of what is feasible to calculate. For example, if we consider a minimal or double-zeta basis set to be sufficient for our needs, we will be able to treat many more atoms with any method than we would if a larger triple-zeta basis set with diffuse functions were to be needed.

4.2.2.3 **Density functional theory**

Density functional theory (DFT) provides an alternative to traditional ab initio methods (i.e. MP2 and CCSD(T)) for the inclusion of electron correlation. Whereas ab initio methods focus on a solution for the wavefunction, DFT methods solve for the electron density. The derivation of the DFT approach is again beyond the scope of this chapter. Interested readers are referred to standard texts [67, 68, 92, 93]. Although first widely used in computational physics, DFT has found much recent use in chemistry. Similar to ab initio methods, DFT requires a basis set to represent atomic and molecular orbitals, although basis set free methods have been devised [94]. DFT also requires the functional used to represent the expected electron density around an atom. Separate functionals represent the effects of both electron exchange and electron correlation. Thus, they are generally combined into exchange-correlation functionals. A popular early functional combined Slater exchange with the correlation functional of Vosko, Wilk, and Nossair (SVWN). [95]. More accurate functionals were later created that included the electron density gradient, such as the combination of the exchange functional of Becke [96] with the correlation functional of Lee, Yang, and Parr (BLYP) [97]. Now popular is the hybrid B3LYP functional, which includes the exact Hartree-Fock exchange, the exchange functional of Becke, and the correlation functional of Lee, Yang, and Parr in a parameterized fashion [98]. In many cases, B3LYP gives accuracy comparable to that of traditional ab initio methods, such as MP2 [99]. The popularity of DFT stems from the fact that it has a lower basis set dependence than MP2 or higher-order methods, and thus allows a considerable saving in computer time. An added benefit is that basis set requirements for DFT are generally less than they are for traditional methods. If a given accuracy can be obtained for a given system with a smaller basis set, additional savings in computer time are possible.

Most QM methods were initially developed to treat individual molecules or small complexes, not complete ensembles of molecules. As noted above, periodic boundary conditions are used in classical simulations of gas, liquid, and solid systems. This same technique has been applied to QM methods, resulting in both periodic Hartree-Fock (PHF) [100] and DFT methods. The main advantage of these periodic methods is the possibility of more accurately including long-range electrostatic effects (through Ewald techniques) as well as minimizing "edge effects". The PHF method has been used to study a variety of metals and metal oxides, although the lack of analytical derivatives has limited its use. In contrast, periodic DFT methods, which generally utilize a plane-wave basis set (PWDFT), provide forces at very low computational cost. These methods, often referred to as Carr-Parinello Molecular Dynamics [101], which involve highly efficient 3D Fourier transforms in the evaluation of energies and forces, have been used to study many metal and metal oxide catalytic systems [102, 103, 105]. In addition to obtaining minimum-energy geometries, periodic DFT methods have also been coupled with MD schemes to allow the time evolution of systems to be studied.

4.2.2.4 Prediction of experimental data by quantum mechanical methods

In the section on classical methods, a variety of techniques, such as energy minimization, Monte Carlo, and molecular dynamics were presented, that are based on energies and forces provided by force fields. These techniques are also widely used within QM. Thus, it is common to obtain minimum-energy geometries for molecules and complexes using QM methods. IR and Raman spectra can also be obtained. MC and MD calculations can be run with energies and forces derived from QM methods rather than from a force field.

While QM methods are much more computationally expensive than classical methods, they have several advantages. They generally do not require parameterization, and thus can be applied directly to almost any system. QM methods explicitly include changes in charge density (polarization), which is usually neglected in classical studies or must be parameterized for. Obviously, QM methods reproduce the electronic features of molecules, and can thus predict many electronic properties of molecules that are not accessible by force fields. These include orbital energies, HOMO's and LUMO's, dipole and higher-order moments, and electric field gradients. Also important to catalysis, QM methods can model bond-breaking and -forming, and thus chemical reactions. Indeed, the determination of the energies and geometries of all states along a reaction pathway is vital to many aspects of catalysis. In addition, only QM methods can really provide information about the transition states of reactions, which are also very important to catalysis research.

NMR properties

We specifically single-out the QM calculation of NMR properties, as this has played a large part in our recent research and in the success of our collaboration with the Haw group. NMR chemical shifts provide valuable information about molecular structure and bonding [106]. While the determination of reliable shift data has long been solely the domain of the experimentalist, recent improvements in theoretical methodology now allow the routine calculation of magnetic properties, in particular the chemical shift tensor. Early attempts at the theoretical determination of chemical shifts were hampered by computational expense and the problem of gauge dependence common to all magnetic phenomena. The idea of using gauge-including atomic orbitals (GIAOs, also known as London orbitals), which have a formal dependence on magnetic field strength and thereby circumvent the gauge origin problem, was first implemented at the ab initio Hartree-Fock level by Ditchfield in 1974 [107]. In 1990, Wolinski, Hinton, and Pulay reformulated the GIAO approach in terms of analytical derivative techniques [108], thereby making NMR calculations at the restricted Hartree-Fock (RHF) level much more tractable. Reliable theoretical NMR calculations are now possible using a variety of methods, including the IGLO (individual gauge for localized orbitals) [109–111], LORG (localized orbitals, local origin) [112–114], and GIAO [107, 108] approaches. Unfortunately, neglect of electron correlation has been shown to lead to potentially large errors in the calculated isotropic chemical shifts for molecules with triple bonds, such as CO and acetonitrile [115]. Furthermore, it has recently been shown that isotropic chemical shift calculations at the GIAO-RHF level are unreliable for benzenium [116] and related carbenium ions [117].

As pointed out in the literature [118], electron correlation can enter the calculation of the NMR data in two ways; through the indirect effects of electron correlation on the geometry of the molecules under study, and through the direct effect of inclusion of electron correlation in the calculation of magnetic properties. Gauss recently extended the GIAO approach to include dynamic electron correlation through second-order Møller-Plesset perturbation theory (GIAO-MP2) [119]. Chemical shift calculations at the GIAO-MP2 level with suitably large basis sets significantly improve the calculated isotropic chemical shifts of neutral molecules with triple bonds and the benzenium ion as referenced above. Gauss has since extended the GIAO formalism to the MP3 and SDQ-MP4 levels [120], and with Stanton, to the coupled-cluster singles and doubles (CCSD) level [121]. Methods that include dynamic correlation through density functional theory (DFT) have also been developed. These include the GIAO-DFT [122–124] and SOS-IGLO-DFT [125, 126] methods, as well as IGLO-DFT and LORG-DFT implementations [117]. The effects of static electron correlation with multi-configurational self-consistent field (IGLO-MCSCF [127] and GIAO-MCSCF [128]) wavefunctions have also been studied.

Most theoretical investigations, like experimental applications of NMR, have focused on the determination of isotropic chemical shifts and the interpretation of these data in terms of molecular topology and bonding. However, the familiar isotropic shift is merely one-third of the trace of a second rank chemical shift tensor that contains up to nine unique components. The antisymmetric part of this tensor does not affect the Zeeman energy to first order [129], and only the remaining six components are potential observables in NMR spectra. Three components of the chemical shift tensor can be measured from spectra of randomly oriented powders, and the utility of ^{13}C principal component measurements was apparent from the early work of Waugh and Pines [130]. Experimental determination of all six components of the chemical shift tensor requires that rotation plots of the spectra of a single crystal be obtained in a goniometer probe. Knowledge of the six components allows one to identify the three Euler angles necessary to orient the principal axes of the chemical shift tensor in the molecular coordinate system. While theoretical calculations naturally give all the tensor components, the determination of tensor orientation by experimental methods is less routine. It is obvious that the chemical shift tensor is much more sensitive to structure and bonding than the isotropic value.

In several cases, we have found a linear relationship between isotropic chemical shifts calculated at the GIAO-RHF and GIAO-MP2 levels, although GIAO-MP2 values are closer to experimental values [131]. These results suggest that in some cases trends in chemical shift can be adequately predicted at the RHF level. We have generally found correlations between GIAO-DFT and GIAO-MP2 chemical shifts to be less good, and prefer the RHF results over those from DFT, despite the reasonable assumption that the DFT results should be better as they include electron correlation.

We have found that for our work, which largely involves prediction of ^{13}C and ^1H chemical shifts, the flexibility in the core region of the basis set is important with regard to obtaining converged results. Multiple polarization and diffuse functions often add little accuracy but considerable computational expense. The tzp basis sets of Schäfer and Ahlrichs [132, 133] are very good for our purposes. ^{15}N chemical shifts are much more sensitive to basis set size, and much larger basis sets, and often higher-order correlation treatments, are needed [120]. In some cases, uncontraction of the core region of the basis set leads to significant improvements in accuracy.

4.3
Model Systems

We have already discussed some of the factors that affect of accuracy of theoretical calculations, such as the force field used in classical simulations, and the basis set and electron correlation treatments used in quantum mechanics. Another very important aspect to consider is the actual system to which the theoretical methods are applied. In most experimental studies of catalysis, the results correspond to averages over huge numbers of atoms and molecules, and long periods of time. In practice, classical studies usually treat less than 100,000 atoms, while quantum mechanical studies generally treat less than 1000. As mentioned earlier, even the longest MD simulations cover only a fraction of the time of an experimental measurement. QM methods generally study a single configuration, a limited number of low-energy configurations, or, at best, the time behavior over picoseconds. Thus, theoretical studies are generally forced to treat a smaller model system that hopefully reproduces the experimental behavior.

Our first indications that theoretical studies of smaller model systems could reproduce macroscopic behavior came from classical MC and MD simulations of gases and liquids, typically performed on hundreds to thousands of molecules. These simulations are generally able to reproduce experimental behavior (diffusion constants, phase diagrams, radial distribution functions, etc.) with excellent of accuracy. MC and MD simulations of solid-phase systems, such as zeolites, typically involve less than 50 unit cells with associated adsorbate molecules. Again, these simulations have been able to reproduce and predict much of what we observe experimentally with great faithfulness. Note that these classical simulations generally involve periodic boundary conditions, and thus include long-range interactions. Evidence in the literature thus suggests we can cautiously assume that the models we are using are adequate.

Advances in computer power and methodology have only recently allowed QM studies of an entire zeolite unit cell [134–137]. Thus, many previous studies involved small cluster models that included the area of interest (i.e. a zeolite Brønsted site) and a few surrounding atoms. To satisfy valence, the cluster is terminated, typically with hydrogens. Such clusters models do not include long-rang electrostatic effects, although these may be included by, for example, surrounding the cluster with a set of point charges that hopefully mimic the long-range potential [138]. Despite the limitations of cluster models, they have proven very useful for the study of catalysis, particularly for zeolites. Results in the literature show that for reasonable cluster sizes, various predicted quantities, such as geometries and deprotonation energies of zeolite acid sites, agree with experimental values [139–142]. Thus, many of the chemical effects observed experimentally in zeolites appear to be very local in nature, with long-range interactions playing only a minor role. However, there are clearly cases in which the long-range interactions do seem important, such as the prediction of zeolite ^{17}O and ^{29}Si NMR spectra [143]. Note also that it is much more difficult to construct useful cluster models of particular systems, such as metal surfaces. In these, it is often difficult to obtain the same electronic structure in the cluster as in the true metal (due to the many possible spin states of some metal clusters). As more results from periodic QM and HF methods become available, we will be better able to address this issue [144].

Our common quantum mechanical strategy

For studies of zeolites and zeolite adsorbate systems, we commonly used a cluster model of stoichiometry $(H_3SiO)_3SiOHAl(OSiH_3)_3$, which is derived from the ZSM-5 X-ray crystal structure [145]. The terminal silyl groups are fixed in the crystallographic positions, whereas all other atoms are optimized. We usually optimize the clusters using DFT, the B3LYP hybrid exchange-correlation functional, and the 6-311G* [146, 147] or DZVP2 [148] basis sets. For chemical shift calculations of carbenium ions and other adsorbed species, we optimize at the B3LYP/6-311G* level and calculate the NMR properties at the GIAO-MP2 level, with a tzp basis set for heavy atoms, and dz for hydrogens [132, 133].

4.4
Applications of Theoretical Methods to Catalysis

4.4.1
Diffusion of Adsorbates in Silicalite

Many researchers have used MD simulations to study the diffusion of adsorbates in zeolites. Initial work focused on rare gases and hydrocarbons in silicalite [51, 149–152]. We used MD to study the diffusion of methane and propane in silicalite, using a full valence force field [153]. As part of this work, we tested four different force fields available at the time for the required atoms. After selection of the force field, based on a comparison with experimental heats of adsorption, we were able to make predictions of the diffusion constants for both adsorbates. The simulations were performed with fully flexible adsorbates, whereas the silicalite atoms were frozen. The force field included both Lennard-Jones and electrostatic interactions. The simulations were run at 300 K and utilized 27 unit cells of silicalite with periodic boundary conditions. The results, shown in Fig. 4.1 for a range of loadings, are very close to those of experimental pulsed field gradient NMR measurements, which we feel most closely correspond to the microscopic diffusion constants determined from the simulation. The simulations also correctly predicted the decreases in diffusion constants as the loading of the adsorbates on the zeolite was increased. We later completed simulations of butane and isobutane in silicalite. The predicted diffusion constants are also shown in Fig. 4.1. Butane diffusion is only slightly slower than that of propane, but that of isobutane is almost two orders of magnitude slower. These results clearly illustrate the basis for shape-selective separation by silicalite. For butane and isobutane, we also carried out simulations with a flexible silicalite, using our silicate force field [11]. In both cases, the self-diffusion rates were found to increase by about a factor of three (Fig. 4.1).

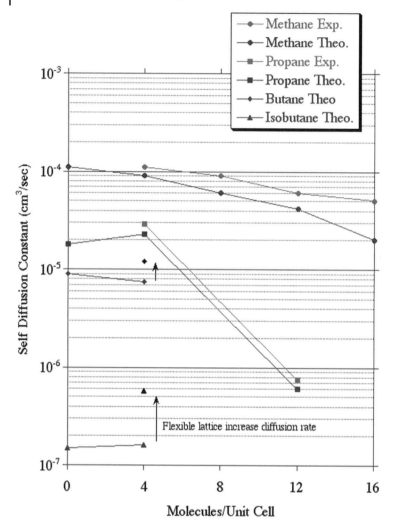

Fig. 4.1 Theoretical and experimental self-diffusion constants for methane, propane, butane, and isobutane as a function of loading. Data for simulations with both a rigid and flexible silicalite lattice are also shown.

4.4.2
Theoretical Characterization of Zeolite Acidity

The exact nature of intermediates in zeolite reactions is well suited for theoretical study. For many years, zeolite hydrocarbon reaction mechanisms were believed to be very similar to those assumed for strong liquid acids, with the invocation of stable, long-lived, carbenium and even carbonium intermediates. One of the first indications that stable carbenium and carbonium ions might not be involved came from the theoretical study of propylene protonation in a zeolite by Kashansky [154]. In this case, rather than the generation of a stable propyl cation, and alternative species, termed a surface alkoxide (probably more correctly an oxonium ion) was assumed to form. This theoretical study complemented the earlier experimental work of Haw and co-workers on the same system [155]. Closely following this publication was work by Kramer and co-workers, which suggested that a stable carbonium ion was not involved in H/D exchange of methane on zeolites [156, 157].

These alternative reaction mechanisms imply that zeolites are weaker acids than commonly supposed, and that they are unable to stabilize cationic intermediates to the extent possible in strong liquid acids. Consistent with this idea, our study of Hammett bases adsorbed on zeolites suggests that zeolite acidity is considerably below the superacid range [158]. In particular, both theoretical and experimental results show that two organic bases with Hammett acidity values at the threshold of superacidity are not protonated in HZSM-5.

A related and controversial aspect of the characterization of zeolite acidity is our recent observation that the only olefinic or aromatic molecules that are observed to form stable carbenium species in unmodified zeolites are also strongly basic [159]. Ab initio calculations at the MP4(sdtq))/6-311+G* level predict that stable carbenium ions form if the parent olefin or aromatic has a proton affinity ≥ 210 kcal mol^{-1}. This work indicates that many hydrocarbons, commonly assumed to form carbenium ions within zeolites, such as benzene and toluene, do not. It is important to point out that although much of our work has cast doubt on the formation of stable carbenium species from weakly basic precursors, both theory and experiment indicate that these three carbenium ions do form within zeolites. The geometries and theoretical proton affinities of the parent compounds, 1-methylindene, 1,3-dimethylcyclopentadiene, and 1,5,6,6-tetramethyl-3-methylenecyclohexa-1,3-diene, are shown in Scheme 4.1

Closely related to the determination of zeolite acid strength is the study of the mechanisms of organic reactions on zeolites. Our study of H/D exchange in benzene combined the first in situ NMR kinetic measurements with a theoretical determination of the reaction mechanism [160]. Both experimental and theoretical results indicate that the reaction proceeds through a benzenium-like transition

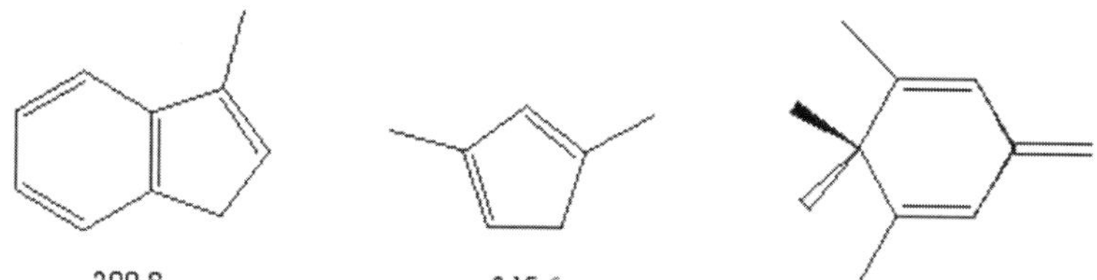

209.8   215.6   227.4

Scheme 4.1

state and not a stable carbenium intermediate. In the van der Waals (VDW) complex, benzene is η^6-coordinated to the acidic zeolite proton with a weak interaction energy of 1.3 kcal mol^{-1} (Fig. 4.2a). The calculated TS geometry (Fig. 4.2b) corresponds to the asymmetric exchange of the "hydrons" (these calculations do not distinguish between isotopes). Although the TS resembles a benzenium cation, the charge on C_6H_7 is only 0.56 | e |. In addition, the bond lengths between the exchanging hydrons and the carbon are ~0.1 Å longer than in a free benzenium cation. More importantly, the TS collapses to the VDW complex in a barrierless transformation, and thus no stable benzenium species is involved. Graphical analysis of the normal mode displacements verifies that an imaginary frequency (−423 cm^{-1}) corresponds to hydron exchange. The theoretical activation energy for H/D exchange, obtained at the B3LYP/6-31G* level, is 20.6 kcal mol^{-1}. The experimental activation energy for benzene H/D exchange on various acidic zeolites ranges be-

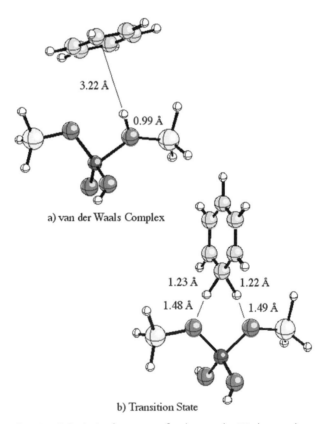

Fig. 4.2 a) Optimized geometry for the van der Waals complex between benzene and the zeolite cluster model. Selected distances in Å. **b)** Optimized geometry for the transition state for H/D exchange.

tween 14.4 and 25.5 kcal mol^{-1}. The experimental value on ultra-stable Y (USY), the zeolite most closely modeled by the cluster in Fig. 4.2, is 20.3 kcal mol^{-1}. The transition state structure in Fig. 4.2b shows a strong resemblance to those determined for H/D exchange for methane by Kramer and van Santen [156, 157], and for hydrogen by Evleth and co-workers [161].

4.4.3
Solvent-Assisted Proton Transfer in Catalysis by Zeolite Solid Acids

We have established that the adsorption of "solvent" molecules within zeolite channels can have a strong effect on the acidity of the zeolite, and thus alter the reaction mechanism and product distribution [162]. Experimental ^{13}C NMR measurements were made at cryogenic temperature and low coverages for acetone in HZSM-5, in the presence of additional co-adsorbates such as nitromethane. Electronic structure calculations employing a continuum dielectric predict that the acidity of the zeolite increases by as much as 35 kcal mol^{-1} as the dielectric constant (ϵ) of the "solvent" increases from 1 to limiting values ($\epsilon \sim 20$). We have also found that co-adsorbed "solvent" molecules result in the common probe molecule acetone being more readily protonated.

Table 4.1 gives the results of calculations of the proton affinity of acetone complexed to six different co-adsorbates. These calculations predict proton affinity increases of as much as 12 kcal mol^{-1} (with acetonitrile) when a co-adsorbate is present [163]. Also included in Table 4.1 are the experimental changes in acetone ^{13}C chemical shift for the different acetone/co-adsorbate systems in HZSM-5. The changes in chemical shift indicate increased proton transfer to acetone as we vary the co-adsorbate from propane to acetonitrile. This is consistent with the predicted increases in acetone proton affinity.

Figure 4.3 shows the optimized geometry of acetone complexed to a zeolite acid site. Figure 4.4 also shows acetone and the zeolite, but in this case with further coordination by nitromethane, which is present as a co-adsorbed solvent molecule. A few key interatomic distances are shown in each figure. Calculations predict that

Co-adsorbate	ΔH	$\Delta\Delta H$	$\Delta\Delta\delta_{iso}$
Acetone	−195.32[a]	0.0	0.0
Propane	−195.89	−0.57	1.2
Bromomethane	−200.34	−5.59	2.2
Nitromethane	−204.73	−9.41	6.8
Nitroethane	−206.05	−10.73	3.7
Nitropropane	−206.40	−11.08	2.6
Acetonitrile	−207.39	−12.07	4.7

[a] **Experimental value 196.7 kcal mol^{-1} [189].**

Tab. 4.1 Proton affinities (ΔH) for acetone and acetone/co-adsorbate complexes (kcal mol^{-1}), changes in ΔH, and the experimental ^{13}C isotropic chemical shift (ppm) relative to acetone.

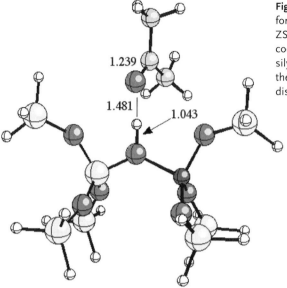

Fig. 4.3 Optimized geometry for acetone complexed to a ZSM-5 model. The coordinates of the terminal silyl groups were held fixed in the crystal geometry. Selected distances in Å.

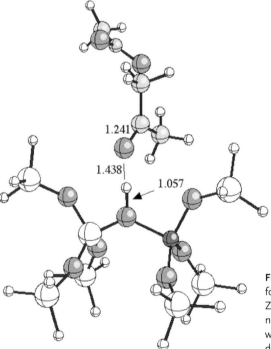

Fig. 4.4 Optimized geometry for acetone complexed to a ZSM-5 model, with nitromethane also in contact with acetone. Selected distances in Å.

with nitromethane present, the O–H distance increases from 1.043 Å to 1.057 Å, while the H··O distance decreases from 1.481 Å to 1.438 Å. There is also a small decrease in the C=O distance. These changes in geometry are consistent with an increased proton transfer from the zeolite to acetone, and effectively, with increased zeolite acidity and acetone basicity.

4.4.4
Activation of Brønsted Acids by Lewis Acids: The Creation of New Solid Acid Catalysts

One way of describing the active site in a zeolite is as a weak Brønsted acid (a silanol group) activated by a Lewis acid (the neighboring Al^{3+} coordinated to framework oxygen). Thus, the bond between a bridging hydroxyl group and Al is dative, involving coordination of the lone pairs of the hydroxyl oxygen to the empty orbital on Al. We used this Brønsted–Lewis activation as a design criterion for new acid catalysts. Inspiration in this area came from our work with Dr. Russell Drago (University of Florida, deceased) on his unique $(SG)_n$-$AlCl_2$ catalyst [164], which is created by the reaction of $AlCl_3$ with the silanol groups on the surface of silica gel. The reaction releases HCl, and creates a new type of active site in which the $AlCl_2$ activates a silanol to from a strong Brønsted acid (Scheme 4.2).

Our combined theoretical and experimental investigation of this catalyst indicated that it was a stronger acid than zeolites, bordering on superacidity [165]. The theoretical work identified factors affecting acidity and predicted the likely geometry of the active site. Figure 4.5 shows the optimized geometry of one of the theoretical models that we created in order to study the acidity of the active site.

Using this result as a design criteria, we tested the acidity of several Lewis acid/Brønsted acid models. We thus created a new solid acid by the activation of silica silanols with SO_3 [166]. Our theoretical work predicts that after coordination of the SO_3 to surface silanols in a dative fashion, further reaction involving proton transfer gives the silyl ester of sulfuric acid grafted on the surface, a thermodynamically more stable arrangement (Scheme 4.3).

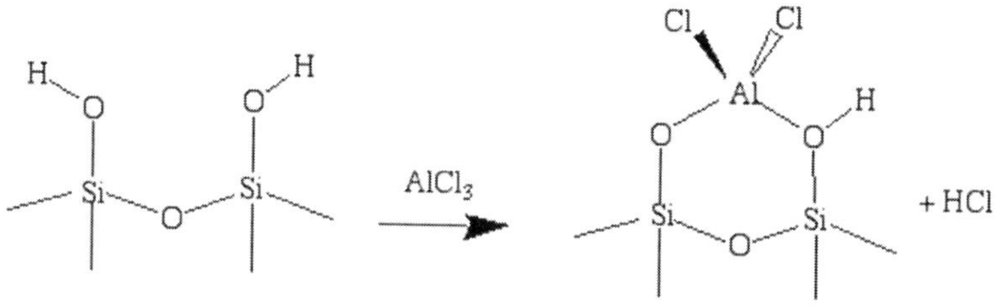

Scheme 4.2

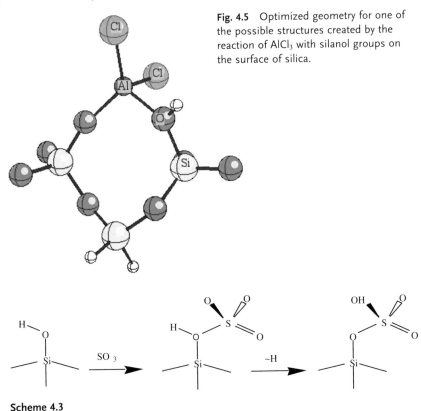

Fig. 4.5 Optimized geometry for one of the possible structures created by the reaction of AlCl₃ with silanol groups on the surface of silica.

Scheme 4.3

This sulfuric acid ester is theoretically predicted to have an acidity very similar to that of H-ZSM5, in agreement with the experimental results. We are currently patenting the process for creating these new catalysts.

4.4.5
Carbenium Ion Chemistry on Solid Acids: Theoretical NMR

Considering that we have spent a great deal of effort in disproving commonly-held beliefs with regard to carbenium ion formation in zeolites, it is also important to prove that such species can be created and observed under suitable conditions. We have found that Lewis acid powders have much greater acidity than typical zeolites, and therefore readily generate carbenium species. As many of these carbenium ions have unusual NMR chemical shifts, it is important to be able to theoretically verify the identity of these species.

Our first use of the GIAO-MP2 method involved the identification of the isopropyl cation, formed on frozen SbF₅ [167]. Although the agreement between the theoretical and experimental ¹³C NMR was close for even the bare cation, we found that including a charge-balancing anion improved the agreement. The optimized geometry of the isopropyl cation coordinated to an SbF_6^- counterion if shown in Fig. 4.6. We were also able to obtain the orientation of the chemical shift tensor, which we could not obtain experimentally.

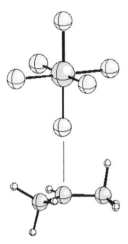

Fig. 4.6 Optimized geometry for the isopropyl cation complexed to an SbF_5^- counter ion. The C–F distance was held fixed during the optimization.

We later expanded on this approach in the study of seven oxocarbenium ions, also created on Lewis acid powders [168]. The agreement between the theoretical and experimental ^{13}C NMR isotropic shifts and tensor components, shown in Table 4.2, is excellent. In this case, we did not find it necessary to include charge-balancing counterions in order to achieve a close match to the experimental data. While agreement without the counterions could have been considered fortuitous in a single case, the fact that seven oxocarbenium ions could be modeled in this way suggests the Lewis acid catalyst has little effect on the ^{13}C NMR data.

The presence of the 1,2,2,3,5-pentamethylbenzenium cation in HZSM-5 was verified by GIAO-MP2 calculations [169]. Also verified by theoretical methods were the 1,3-dimethylcyclopentadienyl and 1,2,3-trimethylcyclopentadienyl cations [170]. GIAO-MP2 calculations were also used to distinguish between the 1,1,2,2,3,4,5-heptamethylcyclopentadienyl cation and the 1,1,2,3,4,5,5-heptamethylcyclohexadienyl cation in a study of methanol-to-olefin chemistry on HSAPO-34

		Theoretical–Experimental (ppm)			
R-	δ_{iso} (exp.)	δ_{iso}	δ_{11}	δ_{22}	δ_{33}
CH_3-	152	0	2	2	−9
CH_3CH_2-	154	0	−8	−30	−17
$(CH_3)_2CH$-	155	0	−4	−19	−12
$(CH_3)_3C$-	155	0	5	5	−5
$ClCH_2$-	146	2	−30	41	−5
C_6H_5-	154	3	1	10	0
FC_6H_4-	155	0	6	43	−46

Tab. 4.2 Experimental ^{13}C isotropic chemical shifts, and differences between theoretical and experimental values for the isotropic shifts and the tensor components (ppm) for the carbenium carbon. R–C^+≡O is the parent oxocarbenium ion.

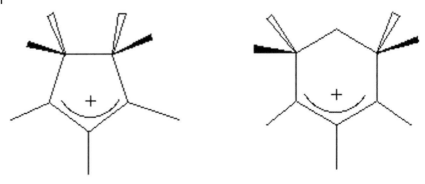

Scheme 4.4

(Scheme 4.4) [171]. We are generally able to attain agreement to within ~2 ppm for the ^{13}C chemical shifts of the carbenium ions we study.

In many cases, the systems for which we would like to calculate NMR data are too large to be treated by the GIAO-MP2 method. In a theoretical study of acetone adsorbed on a wide range of Brønsted and Lewis acids, we found that linear regression could be used to estimate MP2 values from RHF values [131]. We were thus able to accurately predict isotropic chemical shifts for large systems such as acetone adsorbed on an H-ZSM5 model, and acetone adsorbed on an AlCl$_3$ dimer. As a further test of the theoretical methods, we showed that electron correlation, generally assumed to be unimportant for the calculation of ^{1}H NMR data, has a large effect on the predicted chemical shifts if the protons are involved in strong hydrogen bonds [172].

4.4.6
Base Catalysis by Metal Oxide Surfaces

Building on the success of our use of the combined theoretical and experimental approach in the study of solid acid catalysis, we applied the same methodology to the study of basic catalysts. Our first work in this area involved the study of acetylene adsorption on MgO nanoparticles [173]. Using a variety of models of the MgO surface, we were able to verify that acetylene reacts to form an acetylide at low coordination sites. The optimized geometry of the acetylide on our largest MgO model is shown in Fig. 4.7a. In accordance with the results of other studies of adsorption on surfaces, the acetylene preferentially interacts with three-coordinate sites; four-coordinate sites are less favored, while we were unable to predict any binding to five-coordinate sites. The GIAO predictions of the ^{13}C NMR spectra of the acetylide were in excellent agreement with the experimental data. The alternative possibility, i.e. the formation of a vinylidene, was ruled out. We were also able to identify an ethoxy species that forms on the aged catalyst. The optimized geometries of two such ethoxy species are shown in Figs. 4.7b and 4.7c. The difference in predicted chemical shifts between these different ethoxy species was not sufficiently large to distinguish which ethoxy group was being observed. Further verification of all adsorbed species was provided by a comparison of theoretical and experimental IR spectra.

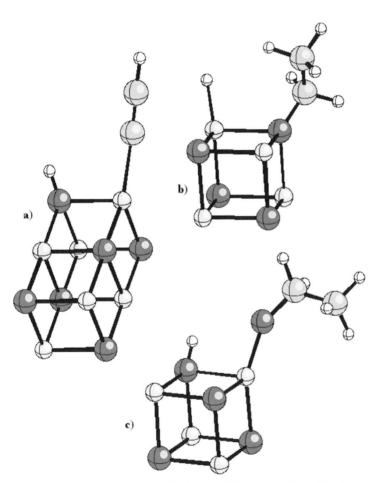

Fig. 4.7 **a)** Optimized geometry for the acetylide complex formed by the reaction of acetylene with a three-coordinate site on MgO. **b)** Optimized geometry for one of the ethoxy complexes possibly formed with three-coordinate sites on MgO. **c)** Optimized geometry for another possible ethoxy complex. In this case, we assume that a surface hydroxyl group is also involved.

4.5
Future Developments in Computational Catalysis

4.5.1
Computing Power

Even in the time it takes for this chapter to appear in press, there will be additional advances in computing. Computers will become faster and cheaper. For many years, the computer power available for a given price has doubled about every 18 months. Theoretical chemists will be obvious beneficiaries of this trend. In addition, the many researchers working on new algorithms will doubtless provide us with new, faster, and more accurate methods. We believe that this huge increase in capability is unparalleled in any other aspect of catalysis research. Theoretical chemistry can only become more valuable in the study of catalysis.

A major force in computing is the increasing popularity of Linux, a freely available operating system [174]. Many academic and commercial codes have already been ported to Linux. Many groups have been able to create their own "supercomputers" by linking commodity personal computer parts with fast networks such as Giganet or Myrinet, and running then with Linux [175]. These Linux or so-called "Beowulf" clusters are already challenging the market now held by large, mainframe-type computers. We expect the migration to Linux and, in particular, to Linux cluster computing to continue for some time. However, it is certainly possible that other commercial operating systems may supplant Linux over a few years.

4.5.2
Approvements in General Methodology

As noted several times above, we expect significant developments in algorithms, and increased use of parallelism. Several research groups are working on lessening the basis set dependence of several QM methods. For example, pseudospectral RHF and DFT methods [176, 177] already allow much larger systems to be studied than traditional computer codes. The local MP2 method (LMP2) [178] and the Resolution of the Identity MP2 (RI-MP2) method [179,180] similarly allow much larger MP2 calculations than conventional methods. We expect continued advances.

4.5.3
Plane-Wave DFT

DFT methods that utilize a plane-wave basis set (PWDFT) can treat periodic systems, allowing a more realistic representation of metal, metal oxide, and other catalytic systems than afforded by cluster models [104, 181, 182]. PWDFT has already been used to study a wide variety of metal catalysts with and without reactant molecules [103–105]. PWDFT has also been used to study the behavior of molecules adsorbed in zeolites [181]. The plane wave basis set lends itself to efficient evaluation and parallelization. We expect the use of PWDFT methods to become more widespread.

4.5.4
Combinatorial Catalysis

Combinatorial chemistry and high-throughput screening (HTS) have revolutionized the discovery of pharmaceuticals. These techniques are now being applied to the development of other biological molecules, such as herbicides and pesticides. Similar techniques are just beginning to be used in materials science, such as for the generation of ligands in homogeneous catalysis and the design of materials with improved magnetic and optical properties [183]. Computational chemistry has the potential to provide significant aid to combinatorial synthesis of catalysts and HTS, particularly if sufficient advances in computational techniques and computer power can be made. Theoretical research in HTS of catalysts could take a variety of directions.

The overall goal of combinatorial methods is to produce large libraries of molecules by automatic synthesis [184], and then to rapidly screen them by similarly automated processes [185]. The hope is that within the large library of molecules, a few may show the desired activity. Theoretical methods have found great use in pharmaceutical chemistry as much can be done with classical force field determinations of conformations and similar comparisons, e.g. to test whether the molecules will interact favorably with a target enzyme, or even show similarity to other known active compounds. These calculations, which typically only consider molecular volume, orientation of functional groups, and hydrogen bonding, can often take only a second or less of CPU time per molecule, allowing thousands, millions, and even billions of molecules to be screened.

The utility of theoretical chemistry in a combinatorial approach is much less clear. Although some aspects of catalysis may depend strongly on steric interactions, such as the well know Tolman cone angle [186] or the stereoselectivity of olefin polymerization catalysts [187], many aspects of catalyst design require electronic structure calculations. Even with current hardware and software advances, calculating the electronic structure of even 100's of catalytic systems is a huge task. Although we may envisage more use of the combinatorial approach in a few years from now, much will also depend on the popularity of the combinatorial approach within the experimental catalysis community [188].

Acknowledgments

This work was supported by the National Science Foundation (CHE-9996109) and the US Department of Energy. Computer resources were provided by the National Energy Research Supercomputer Center (NERSC), Berkeley, CA.

References

1 Haw, J. F.; Nicholas, J. B.; Xu, T.; Beck, L. W.; Ferguson, D. B., *Acc. Chem. Res. 1996, **29**, 259–267.*

2 Burkert, U.; Allinger, N. L., *Molecular Mechanics* (Ed.: Caserio, M. C.); American Chemical Society: Washington D. C., 1982; pp. 339.

3 Hopfinger, A. J.; Pearlstein, R. A., *J. Comput. Chem. 1984, **5**, 486–499.*

4 Hunenberger, P. H.; Van Gunsteren, W. F., *Comput. Simul. Biomol. Syst. 1997, **3**, 3–82.*

5 Wang, J.; Kollman, P. A., *J. Comput. Chem. 2001, **22**, 1219–1228.*

6 Weiner, S. J.; Kollman, P. A.; Case, D. A.; Singh, U. C.; Ghio, C.; Alagona, G.; Profeta, S., Jr.; Weiner, P., *J. Am. Chem. Soc. 1984, **106**, 765–784.*

7 Brooks, B. R.; Bruccoleri, R. E.; Olafson, B. D.; States, D. J.; Swaminathan, S.; Karplus, M., *J. Comput. Chem. 1983, **4**, 187–217.*

8 Jorgensen, W. L.; Tirado-Rives, J., *J. Am. Chem. Soc. 1988, **110**, 1657–1671.*

9 Keldsen, G. L.; Nicholas, J. B.; Carrado, K. A.; Winans, R. E., *J. Phys. Chem. 1994, **98**, 279.*

10 Stellpflug, A. G.; Nicholas, J. B.; Carrado, K. A.; Keldsen, G. L.; Winans, R. E., *201st ACS National Meeting, 1991.*

11 Nicholas, J. B.; Hopfinger, A. J.; Iton, L. E.; Trouw, F. R., *J. Am. Chem. Soc. 1991, **113**, 4792.*

12 McCarthy, M. I.; Schenter, G. K.; Scamehorn, C. A.; Nicholas, J. B., *J. Phys. Chem. 1996, **100**, 16989–16995.*

13 Kiselev, A. V.; Du, P. Q., *J. Chem. Soc., Faraday Trans. 2 1981, **77**, 1–15.*

14 van Beest, B. W. H.; Kramer, G. J.; van Santen, R. A., *Phys. Rev. Lett. 1990, **64**, 1955–1958.*

15 Grigoras, S.; Lane, T. H., *J. Comput. Chem. 1988, **9**, 25–39.*

16 Leherte, L.; Vercauteren, D. P.; Derouane, E. G.; Lie, G. C.; Clementi, E.; Andre, J.-M., *in Zeolites: Facts, Figures, Future* (Eds.: Jacobs, P. A.; van Santen, R. A.), Elsevier Science Publishers: Amsterdam, 1989; Number of pages in book cannot be determined.

17 Loncharich, R. J.; Brooks, B. R., *Proteins: Structure, Function, and Genetics, 1989, **6**, 32–45.*

18 Chirlian, L. E.; Francl, M. M., *J. Comput. Chem. 1987, **8**, 894.*

19 Ewald, P., *Ann. Phys. 1932, **64**, 253.*

20 Darden, T.; York, D.; Pedersen, L. G., *J. Chem. Phys. 1993, **98**, 10089.*

21 Essmann, U.; Perera, L.; Berkowitz, M. L.; Darden, T.; Lee, H.; Pedersen, L. G., *J. Chem. Phys. 1995, **103**, 8577.*

22 de Boer, K.; Jansen, A. P. J.; van Santen, R. A., *Chem. Phys. Lett. 1994, **223**, 46–53.*

23 de Boer, K.; Jansen, A. P. J.; van Santen, R. A.; Parker, S. C., *Comput. Mater. Sci. 1996, **6**, 319–330.*

24 Bernardo, D. N.; Ding, Y.; Krogh-Jespersen, K.; Levy, R., *J. Phys. Chem. 1994, **98**, 4180–4187.*

25 Ahlstrom, P.; Wallqvist, A.; Engstrom, S.; Jonsson, B., *Mol. Phys. 1989, **68**, 563–581.*

26 Kutteh, R.; Nicholas, J. B., *Comput. Phys. Commun. 1995, **86**, 236–254.*

27 Kutteh, R.; Nicholas, J. B., *Comput. Phys. Commun. 1995*, **86**, 227–235.

28 Rick, S. W.; Stuart, S. J.; Berne, B. J., *J. Chem. Phys. 1994*, **101**, 6141.

29 de Man, A. J. M.; van Beest, B. W. H.; Leslie, M.; van Santen, R. A., *J. Phys. Chem. 1990*, **94**, 2524–2534.

30 Schlick, T., *Rev. Comput. Chem. 1992*, **3**.

31 Leach, A. R., *Rev. Comput. Chem. 1991*, **2**.

32 Mabilia, M.; Pearlstein, R. A.; Hopfinger, A. J., *J. Am. Chem. Soc. 1987*, **109**, 7960–7968.

33 Metropolis, N.; Rosenbluth, A. W.; Rosenbluth, M. N.; Teller, A. H.; Teller, E., *J. Chem. Phys. 1953*, **21**, 1087–1092.

34 Born, M.; Karman, T. V., *Physik. Z. 1912*, **13**, 297–309.

35 No format in JACS style for this reference type: please edit the JACS style.

36 Trout, B. L.; Chakraborty, A. K.; Bell, A. T., *Chem. Eng. Sci. 1997*, **52**, 2265–2276.

37 Maginn, E. J.; Bell, A. T.; Theodorou, D. N., *J. Phys. Chem. 1995*, **99**, 2057–2079.

38 June, R. L.; Bell, A. T.; Theodorou, D. N., *J. Phys. Chem. 1990*, **94**, 1508–1516.

39 Snurr, R. Q.; Bell, A. T.; Theodorou, D. N., *J. Phys. Chem. 1993*, **97**, 13742–13752.

40 Alder, B. J.; Wainwright, T. E., *J. Chem. Phys. 1957*, **27**, 1208–1209.

41 Rahman, A., *Phys. Rev. 1964*, **136**, A405–411.

42 Berendsen, H. J. C.; Postma, J. P. M.; van Gunsteren, W. F.; DiNola, A.; Haak, J. R., *J. Chem. Phys. 1984*, **81**(8), 3684–3690.

43 Nose, S.; Klein, M. L., *Mol. Phys. 1981*, **50**, 1055–1076.

44 Ryckaert, J. P.; Ciccotti, G.; Berendsen, H. J. C., *J. Comput. Phys. 1977*, **23**, 327–341.

45 Barker, J. A.; Fisher, R. A.; Watts, R. O., *Mol. Phys. 1971*, **21**, 657–673.

46 Tidor, B., *J. Phys. Chem. 1993*, **97**, 1069–1073.

47 Fritzsche, S.; Haberlandt, R.; Kärger, J.; Pfeifer, H.; Heinzinger, K., *Chem. Phys. Lett. 1992*, **198**, 283–287.

48 Fritzsche, S.; Haberlandt, R.; Kärger, J.; Pfeifer, H.; Wolfsberg, M., *Chem. Phys. Lett. 1990*, **171**, 109–113.

49 Leherte, L.; Lie, G. C.; Swamy, K. N.; Clementi, E.; Derouane, E. G.; Andre, J. M., *Chem. Phys. Lett. 1988*, **145**, 237–241.

50 Nowak, A. K.; den Ouden, C. J. J.; Pickett, S. D.; Smit, B.; Cheetham, A. K.; Post, M. F. M.; Thomas, J. M., *J. Phys. Chem. 1991*, **95**, 848–854.

51 Pickett, S. D.; Nowak, A. K.; Thomas, J. M.; Peterson, B. K.; Swift, J. F. P.; Cheetham, A. K.; den Ouden, C. J. J.; Smit, B.; Post, M. F. M., *J. Phys. Chem. 1990*, **94**, 1233–1236.

52 Schuring, D.; Jansen, A. P. J.; van Santen, R. A., *J. Phys. Chem. B 2000*, **104**, 941–948.

53 Demontis, P.; Fois, E. S.; Gamba, A.; Manunza, B.; Suffritti, G. B., *J. Mol. Struct. 1983*, **93**, 245–254.

54 Demontis, P.; Suffritti, G. B.; Alberti, A.; Quartieri, S.; Fois, E. S.; Gamba, A. *Gazz. Chim. Ital. 1986*, **116**, 459–466.

55 Demontis, P.; Suffritti, G. B.; Quartieri, S.; Fois, E. S.; Gamba, A., *Zeolites 1987*, **7**, 522–527.

56 Demontis, P.; Suffritti, G. B.; Quartieri, S.; Fois, E. S.; Gamba, A., *J. Phys. Chem. 1988*, **92**, 867–871.

57 Hedman, F.; Laaksonen, A., *Theor. Comput. Chem. 1999*, **7**, 231–280.

58 Berendsen, H. J. C.; van der Spoel, D.; van Drunen, R., *Comput. Phys. Commun. 1995*, **91**, 43–56.

59 Hedman, F.; Laaksonen, A., *Mol. Simul. 1995*, **14**, 235–244.

60 Kale, L.; Skeel, R.; Bhandarkar, M.; Brunner, R.; Gursoy, A.; Krawetz, N.; Phillips, J.; Shinozaki, A.; Varadarajan, K.; Schulten, K., *J. Comput. Phys. 1999*, **151**, 283–312.

61 Plimpton, S. J.; Hendrickson, B. A., *Mater. Res. Soc. Symp. Proc. 1993,* **291**, 37–42.

62 Smith, W.; Forester, T. R., *Mol. Graph. 1996,* **14**, 136.

63 Allen, M. P.; Tildesley, D. J., *Computer Simulation of Liquids;* Oxford University Press: Oxford, 1987; 385 pages; pp. 155–162 .

64 Hirst, D. M., *A Computational Approach to Chemistry;* Blackwell Scientific Publications: Oxford, 1990; Number of pages in book cannot be determined.

65 Leach, A. R., *Molecular Modeling Principles and Applications;* Addison Wesley Longman Limited, 1996; 595 pages.

66 Hinchliffe, A., *Ab Initio Determination of Molecular Properties;* Adam Hilger: Bristol, 1987; pp. 164.

67 Parr, R. G.; Yang, W., *Density-Functional Theory of Atoms and Molecules;* Oxford University Press: New York, 1989; Number of pages in book cannot be determined.

68 Seminario, J. M.; Politzer, P., *Modern Density Functional Theory: A Tool for Chemistry;* Elsevier: New York, 1995; 418 pages.

69 Hehre, W. J.; Radom, L.; Schleyer, P.; Pople, J. A., *Ab Initio Molecular Orbital Theory;* Wiley: New York, 1986; 576 pages.

70 Szabo, A.; Ostlund, N. S., *Modern Quantum Chemistry: Introduction to Advanced Electronic Structure Theory;* MacMillan Publishing Co. Inc.: New York, 1982; 466 pages.

71 Pople, J. A.; Beveridge, D. L., *Approximate Molecular Orbital Theory;* McGraw-Hill: New York, 1970; Number of pages in book cannot be determined.

72 Segal, G. A.; Pople, J. A., *J. Chem. Phys. 1966,* 3289.

73 Pople, J. A.; Beveridge, D.; Dobosh, P., *J. Chem. Phys. 1967,* **47**, 2026.

74 Pople, J. A.; Segal, G. A., *J. Chem. Phys. 1965,* **43**, S136–S149.

75 Bingham, R. C.; Dewar, M. J. S.; Lo, D. H., *J. Am. Chem. Soc. 1975,* **97**, 1285–1293.

76 Dewar, M. J. S.; Theil, W., *J. Am. Chem. Soc. 1977,* **99**, 4899–4907.

77 Dewar, M. J. S.; Zoebisch, E. G.; Healy, E. F., *J. Am. Chem. Soc. 1985,* **107**, 3902.

78 Stewart, J. J. P., *J. Comput. Chem. 1989,* **10**, 221.

79 Stewart, J. J. P., *J. Comput. Chem. 1989,* **10**, 209.

80 Anderson, W. P.; Edwards, W. D.; Zerner, M. C., *J. Inorg. Chem. 1986,* **25**, 2728.

81 Zerner, M. C.; Loew, G.; Kirchner, R.; Mueller-Westerhoff, U., *J. Am. Chem. Soc. 1980,* **102**, 589.

82 Thiel, W.; Voityuk, A. A., *J. Mol. Struct. (Theochem) 1994,* **313**, 141–154.

83 Mulliken, R. S., *J. Chem. Phys. 1955,* **23**, 1833–1840.

84 Mulliken, R. S., *J. Chem. Phys. 1955,* **23**, 1840–1846.

85 Bartlett, R. J., *J. Phys. Chem. 1989,* **93**, 1697.

86 Pople, J. A.; Seeger, R.; Krishan, R., *Int. J. Quant. Chem. Symp. 1977,* **11**, 149.

87 Hehre, W. J.; Ditchfield, R.; Pople, J. A., *J. Chem. Phys. 1972,* **56**, 2257.

88 Gordon, M. S., *Chem. Phys. Lett. 1980,* **76**, 163.

89 Hariharan, P. C.; Pople, J. A., *Theor. Chim. Acta 1973,* **28**, 213.

90 Dunning, T. H., Jr., *J. Chem. Phys. 1989,* **90**, 1007–1023.

91 Kendall, R. A.; Dunning, T. H., Jr; Harrison, R. A., *J. Chem. Phys. 1992,* **96**, 6796–6806.

92 Chemistry, Department of Theoretical Chemistry. *Amsterdam Density Functional (ADF), Users Guide, Release 1.1;* Free University Amsterdam, 1994.

93 Labanowski, J., Andzelm, J. *J. Density Functional Methods in Chemistry;* Springer-Verlag, New York, 1991.

94 Becke, A. D., *Int. J. Quant. Chem. 1989,* **S23**, 599.

95 Vosko, S. J.; Wilk, L.; Nussair, M., *Can. J. Phys. 1980,* **58**, 1200–1211.

96 Becke, A. D., *Phys. Rev. A 1988,* **38**, 3098–3100.

97 Lee, C.; Yang, W.; Parr, R. G., *Phys. Rev. B 1988,* **37**, 786.

98 Becke, A. D., *J. Chem. Phys. 1993,* **98**, 5648.

99 Johnson, B. G.; Gill, P. M.; Pople, J. A., *J. Chem. Phys. 1993,* **88**, 5612.

100 Pisani, C.; Dovesi, R.; Roetti, C., *Hartree-Fock Ab Initio Treatment of Crystalline Systems;* Springer-Verlag; New York, 1988; Number of pages in book cannot be determined.

101 Carr, R.; Parinello, M., *Phys. Rev. Lett. 1985,* **55**, 2471–2474.

102 Haase, F.; Sauer, J., *Microporous Mesoporous Mater. 2000,* **35–36**, 379–385.

103 Pallassana, V.; Neurock, M., *J. Phys. Chem. B 2000,* **104**, 9449–9459.

104 Desai, S. K.; Pallassana, V.; Neurock, M., *J. Phys. Chem. B 2001,* **105**, 9171–9182.

105 Neurock, M.; Zhang, X.; Olken, M.; Jones, M.; Hickman, D.; Calverley, T.; Gulotty, R., *J. Phys. Chem. B 2001,* **105**, 1562–1572.

106 Ernst, R. R.; Bodenhausen, G.; Wokaun, A., *Principles of Magnetic Resonance in One and Two Dimensions* (Eds.: Breslow, R.; Goodenough, J. B.; Halpern, J. Rowlinson, J. S.); Oxford University Press, 1987; 640 pages.

107 Ditchfield, R., *Mol. Phys. 1974,* **27**, 789–807.

108 Wolinski, K.; Hinton, J. F.; Pulay, P., *J. Am. Chem. Soc. 1990,* **112**, 8251–8260.

109 Schindler, M.; Kuttzelnigg, W., *J. Chem. Phys. 1982,* **76**, 1919–1933.

110 Kutzelnigg, W., *Isr. J. Chem. 1980,* **19**, 193.

111 Kutzelnigg, W., *J. Mol. Struct. (Theochem) 1989,* **202**, 11–61.

112 Bouman, T. D.; Hansen, A. E., *Chem. Phys. Lett. 1990,* **175**, 292.

113 Hansen, A. E.; Bouman, T. D., *J. Chem. Phys. 1985,* **82**, 5035–5047.

114 Hansen, A. E.; Bouman, T. D., *J. Chem. Phys. 1989,* **91**, 3552–3560.

115 Gauss, J., *J. Chem. Phys. 1993,* **99**, 3629–3643.

116 Sieber, S.; Schleyer, P. v. R.; Gauss, J., *J. Am. Chem. Soc. 1993,* **115**, 6987–6988.

117 Arduengo, J. A.; Dixon, D. A.; Kumashiro, K. K.; Lee, C.; Power, W. P.; Zilm, K. W., *J. Am. Chem. Soc. 1994,* **116**, 6361–6367.

118 Bühl, M.; Gauss, J.; Hofmann, M.; Schleyer, P. v. R., *J. Am. Chem. Soc. 1993,* **115**, 12385–12390.

119 Gauss, J., *Chem. Phys. Lett. 1992,* **191**, 614–620.

120 Gauss, J., *Chem. Phys. Lett. 1994,* **229**, 198–203.

121 Gauss, J.; Stanton, J. F., *J. Chem. Phys. 1995,* **103**, 3561–3577.

122 Lee, A. M.; Handy, C.; Cowell, S. M., *J. Chem. Phys. 1995,* **103**, 10095–10109.

123 Rauhut, G.; Puyear, S.; Wolinski, K.; Pulay, P., *J. Phys. Chem. 1996,* **100**, 6310–6316.

124 Cheeseman, J. R.; Trucks, G. W.; Keith, T. A.; Frisch, M. J., *J. Chem. Phys. 1996,* **104**, 5497–5509.

125 Malkin, V. G.; Malkina, O. L.; Casida, M. E.; Salahub, D. R., *J. Am. Chem. Soc. 1994,* **116**, 5898–5908.

126 Schreckenbach, G.; Ziegler, T., *J. Phys. Chem. 1995,* **99**, 606–611.

127 van Wullen, C.; Kutzelnigg, W., *Chem. Phys. Lett. 1993,* **205**, 563–571.

128 Ruud, K.; Helgaker, T.; Kobayashi, R.; Jorgensen, P.; Bak, K. L.; Jensen, H. J. A., *J. Chem. Phys. 1994,* **100**, 8178–8185.

129 Haeberlen, U., *High-Resolution NMR in Solids;* Academic Press: New York, 1976; Number of pages in book cannot be determined.

130 Pines, A.; Gibby, M. G.; Waugh, J. S., *Chem. Phys. Lett.* 1972, **15**, 373–376.

131 Barich, D. H.; Nicholas, J. B.; Xu, T.; Haw, J. F., *J. Am. Chem. Soc.* 1998, **120**, 12342–12350.

132 Schäfer, A.; Huber, C.; Ahlrichs, R., *J. Chem. Phys.* 1992, **100**, 5829–5835.

133 Schäfer, A.; Horn, H.; Ahlrichs, R., *J. Chem. Phys.* 1992, **97**, 2571–2577.

134 Nicholas, J. B.; Hess, A. C., *J. Am. Chem. Soc.* 1994, **116**, 5428.

135 White, J. C.; Nicholas, J. B.; Hess, A. C., *J. Phys. Chem.* 1997, **101**, 590–595.

136 White, J. C.; Hess, A. C., *J. Phys. Chem.* 1993, **97**, 6398.

137 White, J. C.; Hess, A. C., *J. Phys. Chem.* 1993, **97**, 8703.

138 Allavena, M.; Seti, K.; Kassab, E.; Ferenczy, G.; Angyan, J. G., *Chem. Phys. Lett.* 1990, **168**, 461–467.

139 Stave, M. S.; Nicholas, J. B., *J. Phys. Chem.* 1995, **99**, 15046–15061.

140 Brand, H. V.; Curtiss, L. A.; Iton, L. E., *J. Phys. Chem.* 1992, **96**, 7725–7732.

141 Brand, H. V.; Curtiss, L. A.; Iton, L. E., *J. Phys. Chem.* 1993, **97**, 12773–12782.

142 Kramer, G. J.; de Man, A. J. M.; van Santen, R. A., *J. Am. Chem. Soc.* 1991, **113**, 6435–6441.

143 Bull, L. M.; Bussemer, B.; Anupold, T.; Reinhold, A.; Samoson, A.; Sauer, J.; Cheetham, A. K.; Dupree, R., *J. Am. Chem. Soc.* 2000, **122**, 4948–4958.

144 Brandle, M.; Sauer, J.; Dovesi, R.; Harrison, N. M., *J. Chem. Phys.* 1998, **109**, 10379–10389.

145 van Koningsveld, H.; van Bekkum, H.; Jansen, J. C., *Acta Crystallogr.* 1987, **B43**, 127–132.

146 McLean, A. D.; Chandler, G. S., *J. Chem. Phys.* 1980, **72**, 5639–5648.

147 Krishan, R.; Binkley, J. S.; Seeger, R.; Pople, J. A., *J. Chem. Phys.* 1980, **72**, 650.

148 Godbout, N.; Salahub, D. R.; Andzelm, J.; Wimmer, E., *Can. J. Chem.* 1992, **70**, 560–571.

149 June, R. L.; Bell, A. T.; Theodorou, D. N., *J. Phys. Chem.* 1990, **94**, 8232–8240.

150 June, R. L.; Bell, A. T.; Theodorou, D. N., *J. Phys. Chem.* 1990, **94**, 1508–1516.

151 Demontis, P.; Fois, E. S.; Suffritti, G. B., *J. Phys. Chem.* 1990, **94**, 4329–4334.

152 Cohen De Lara, E.; Kahn, R.; Goulay, A. M., *J. Chem. Phys* 1989, **90**, 1485–End page unknown.

153 Nicholas, J. B.; Trouw, F. R.; Mertz, J. E.; Iton, L. E.; Hopfinger, A. J., *J. Phys. Chem.* 1993, **97**, 4149.

154 Kazansky, V. B., Acc. Chem. Res. 1991, **24**, 379–383.

155 Haw, J. F.; Richardson, B. R.; Oshiro, I. S.; Lazo, N. D.; Speed, J. A., *J. Am. Chem. Soc.* 1989, **111**, 2052–2058.

156 Kramer, G. J.; van Santen, R. A., *J. Am. Chem. Soc.* 1995, **117**, 1766–1776.

157 Kramer, G. J.; van Santen, R. A.; Emeis, C. A.; Nowak, A. K., *Nature* 1993, **363**, 529–531.

158 Nicholas, J. B.; Haw, J. F.; Beck, L. W.; Krawietz, T. W.; Ferguson, D. B., *J. Am. Chem. Soc.* 1995, **117**, pp. 12350–12351.

159 Nicholas, J. B.; Haw, J. F., *J. Am. Chem. Soc.* 1998, **120**, 11804–11805.

160 Beck, L. W.; Xu, T.; Nicholas, J. B.; Haw, J. F., *J. Am. Chem. Soc.* 1995, **117**, 11594–11595.

161 Evleth, E. M.; Kassab, E.; Sierra, L. R., *J. Chem. Phys.* 1994, **98**, 1421–1426.

162 Haw, J. F.; Xu, T.; Nicholas, J. B.; Goguen, P. W., *Nature* 1997, **389**, 332–335.

163 Nicholas, J. B., *Top. Catal.* 1999, **9**, 181–189.

164 Drago, R. S.; Petrosius, S. C.; Kaufman, P. B., *J. Mol. Catal.* **1994**, **89**, 317–328.

165 Xu, T.; Kob, N.; Drago, R. S.; Nicholas, J. B.; Haw, J. F., *J. Am. Chem. Soc.* **1997**, **119**, 12231–12239.

166 Zhang, J.; Nicholas, J. B.; Haw, J. F., *Angew Chem. Int. Ed.* **2000**, **39**, 3302–3304.

167 Nicholas, J. B.; Xu, T.; Barich, D.; Torres, P.; Haw, J. F., *J. Am. Chem. Soc.* **1996**, **118**, 4202–4203.

168 Xu, T.; Torres, P. D.; Barich, D. H.; Nicholas, J. B.; Haw, J. F., *J. Am. Chem. Soc.* **1997**, **119**, 396–405.

169 Xu, T.; Barich, D. H.; Goguen, P. W.; Song, W.; Wang, Z.; Nicholas, J. B.; Haw, J. F., *J. Am. Chem. Soc.* **1998**, **120**, 4025–4026.

170 Haw, J. F.; Nicholas, J. B.; Song, W.; Deng, F.; Wang, Z.; Xu, T.; Heneghan, C. S., *J. Am. Chem. Soc.* **2000**, **122**, 4763–4775.

171 Song, W.; Nicholas, J. B.; Haw, J. F., *J. Phys. Chem. B* **2001**, **105**, 4317–4323.

172 Barich, D. H.; Nicholas, J. B.; Haw, J. F., *J. Phys. Chem.* **2001**, **105**, 4708–4715.

173 Nicholas, J. B.; Khier, A.; Xu, T.; Krawietz, T. R.; Haw, J. F., *J. Am. Chem. Soc.* **1998**, **120**, 10471–10481.

174 Petron, E., *Linux Essential Reference;* New Riders: Indianapolis, IN, 2000; pp. 332.

175 Spector, D. H. M., *Building Linux Clusters;* O'Reilly and Associates, Inc.: Sebastopol, CA, 2000; pp. 332.

176 Ringnalda, M. N.; Belhadj, M.; Friesner, R. A., *J. Chem. Phys.* **1990**, **93**, 3397–3407.

177 Won, Y.; Lee, J. G.; Ringnalda, M. N.; Friesner, R. A., *J. Chem. Phys.* **1991**, **94**, 8152–8157.

178 Beachy, M. D.; Chasman, D.; Friesner, R. A.; Murphy, R. B., *J. Comput. Chem.* **1998**, **19**, 1030–1038.

179 Feyereisen, M.; Fitzgerald, G.; Komornicki, A., *Chem. Phys. Lett.* **1993**, **208**, 359–363.

180 Bernholdt, D. E., *Book of Abstracts,* 215th ACS National Meeting, Dallas, March 29–April 2, 1998, HYS-266.

181 Rozanska, X.; van Santen, R. A.; Hutschka, F.; Hafner, J., *J. Am. Chem. Soc.* **2001**, **123**, 7655–7667.

182 Pallassana, V.; Neurock, M., *J. Phys. Chem. B* **2000**, **104**, 9449–9459.

183 Jandeleit, B.; Weinberg, W. H., *Chem. Ind. (London)* **1998**, 795–798.

184 Weinmann, H., *Nachr. Chem.* **2001**, **49**, 150–154.

185 Willson, R. C.; Hill, D. R.; Gibbs, P. R., *High-Throughput Synth.* **2001**, 271–281.

186 Tolman, C. A., *Chem. Rev.* **1977**, **77**, 313–348.

187 Natta, G.; Pino, P.; Mantica, E.; Danusso, F.; Mazzanti, G.; Peraldo, M., *Chim. Ind. (Milan)* **1956**, **38**, 124–127.

188 Senkan, S., *Angew. Chem. Int. Ed.* **2001**, **40**, 312–329.

189 NIST Webbook, www.nist.gov

5
In Situ Ultraviolet Raman Spectroscopy

Peter C. Stair

Abstract

Ultraviolet excited Raman spectroscopy has emerged as a powerful tool for the in situ chemical and structural analysis of materials and chemical reactions. In this chapter, the practice of UV Raman spectroscopy in the author's laboratory is described, including a description of innovative apparatus for in situ measurements. A sampling of results from two applications, catalysis and tribology, is used to highlight the capabilities of this technique.

5.1
Introduction

Raman spectroscopy is one of the most powerful techniques for the characterization of solids and surfaces of technological importance. The high resolution of Raman spectroscopy ($\sim$1 cm^{-1}) and its wide spectral range (50–5000 cm^{-1}) allow examination of the nature of molecular species [1, 2], identification of crystalline solid phases, and determination of the structure of non-crystalline surface phases [3]. The application of Raman spectroscopy to heterogeneous catalysis has been discussed in a number of excellent reviews [4–8] and a recent book [9]. Many measurements on catalysts have focused on the oxides of molybdenum, tungsten, and vanadium supported on alumina in an effort to identify the surface species formed during catalyst preparation and pretreatment. The active oxide often exists as a thin, surface phase on the supporting oxide, which cannot be detected by X-ray diffraction. Moreover, the relevant spectral region for identification and structure determination of the catalytic oxide phase is 500–1000 cm^{-1}, where strong absorption by the support material makes infrared measurements difficult. Typical oxide supports are relatively weak Raman scatterers, which is an advantage for obtaining spectra selectively from the catalytically active phase. Raman spectroscopy has been successfully used to establish the nature of the oxide precursors both in solution and adsorbed on the support surface, to characterize the oxide phases formed during catalyst activation, and to determine the structure of surface oxide phases that have no three-dimensional, crystalline counterpart [10].

Raman spectroscopy is especially powerful as a tool for in situ experiments under practical reaction conditions. It is one of the few instrumental methods that can provide information about both the solid catalyst and the molecular reagents in a single measurement. It is also potentially useful as a tool to investigate the surface chemistry of lubricants or of chemical vapor deposition. However, the number of reports in the literature in which Raman spectroscopy has been used to study chemistry under realistic reaction conditions is surprisingly small (see, for example, [9]) The application of Raman spectroscopy to catalyst characterization has been limited (1) by the small Raman scattering cross-sections (10^{-28} cm^2) and (2) by interference from sample fluorescence and luminescence, which often produce a huge background compared to the weak Raman bands. In many of the catalytic reactions of interest, the catalysts become covered by carbonaceous residues (coke) during the course of the reaction. These samples often exhibit strong fluorescence when the Raman scattering is excited by a visible wavelength laser. Many catalysts also contain impurities that produce strong fluorescences. This has been particularly true of zeolite-based materials. The fluorescence intensity can be 10^6 times larger than the Raman intensity, making the Raman spectrum undetectable [11]. Consequently, a great deal of effort has been devoted to developing methods for avoiding or minimizing sample fluorescence (see, for example, [12]). One popular method is to excite the Raman scattering at longer wavelengths, for example, by using near-infrared radiation as in the FT-Raman technique. This reduces fluorescence but at the expense of Raman scattering intensity. This approach does not provide a general method suitable for many practical catalytic materials under realistic reaction conditions. For example, many catalysts are intensely colored as a result of coke deposition and fluoresce strongly even in the near-infrared [12]. Moreover, FT-Raman measurements using near-IR excitation on samples at elevated temperature are difficult due to the background arising from sample luminescence.

Sample fluorescence can also be avoided by using an ultraviolet laser to excite Raman scattering. This was first demonstrated in 1984, when Asher and Johnson published a paper in which they showed that fluorescence interference was minimized for a broad range of polycyclic aromatic hydrocarbons by excitation at ultraviolet wavelengths below 260 nm [11]. The physical origin of this surprising result is the rapid rate of internal conversion from the high-energy electronic states excited below 260 nm to low-energy singlet or triplet states that fluoresce at wavelengths above 300 nm [13]. Indeed, many of the samples examined in the author's laboratory produce copious amounts of visible fluorescence that does not interfere with the measurement of the Raman spectrum.

Spectra from a wide variety of catalyst samples have been measured in the author's laboratory using the ultraviolet Raman technique. Initially, the focus of our efforts was simply to investigate the range of sample types for which this method successfully avoids fluorescence. An early example is shown in Fig. 5.1, where a spectrum measured from a commercial 1 wt% Rh/Al_2O_3 catalyst coked at 500°C in naphtha (courtesy of Amoco Oil Company) using conventional visible excitation at 514.5 nm (lower panel) is compared to that obtained by ultraviolet excitation at 257 nm (upper panel) [14]. The spectrum obtained using visible excitation is completely dominated by fluorescence; Raman scattering peaks are undetectable. With 257 nm excitation and signal averaging for 10 minutes, the spectrum is dominated by the Raman bands of coke on the catalyst. Similar results have been obtained for catalytic samples ranging from pure zeolites to coked cracking catalysts [15]. Non-catalytic samples such as diamond films, butyl rubber, boron nitride hard coatings, and even sliding lubricated contacts have also produced beautiful Raman spectra [14, 16].

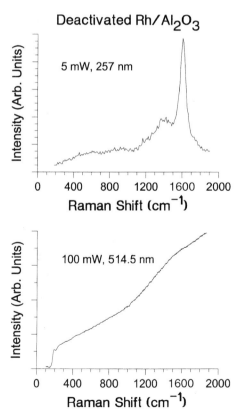

Fig. 5.1 Comparison of Raman spectra from a deactivated, coked rhodium on alumina catalyst measured using visible wavelength excitation (lower) and ultraviolet wavelength excitation (upper).

As alluded to above, there have been excellent reviews published on the application of Raman spectroscopy to catalysis. There are also a number of books and reviews on the theoretical description of Raman scattering. The purpose of this chapter is to describe the application of ultraviolet Raman spectroscopy to in situ measurements from the perspective of the author's laboratory at Northwestern University. In the next section, the instrumentation for in situ measurements by UV Raman spectroscopy is described. The following two sections describe the application of in situ UV Raman spectroscopy in studying catalysts and lubricated sliding contacts, respectively.

5.2
Instrumentation and Experimental Methods

5.2.1
The Spectrometer

The instrumentation used to measure ultraviolet excited Raman spectra is not significantly different from that used to obtain conventional visible Raman spectra. A schematic diagram of the spectrometer used in the author's laboratory is shown in Fig. 5.2. Continuous-wave ultraviolet light in the wavelength range 229–257 nm is generated by intracavity frequency doubling of the radiation in an argon ion laser. A power of more than 1 W at 244 and 257 nm is possible from commercially available lasers. A Pellin–Broca prism (or sometimes a diffraction grating) is used to separate the strong ultraviolet laser line from weak plasma lines emitted from the laser. The excitation laser is focused to a spot (50–100 μm in diameter) on the sample. Scattered light is collected by a wide-angle, ellipsoidal mirror and focused into the entrance slit of a triple-grating spectrograph. The barrel-shaped ellipsoidal mirror collects light over all azimuthal angles and the range of polar angles from 40° to 66°. This corresponds to a minimum effective $f/\#$ of 0.22 and a solid angle of collection of 2.3 steradians, which is nearly 35% of the Raman scattered light in the half-space above the sample. The ellipsoidal mirror also provides $f/\#$ matching to the spectrometer. The first two gratings operate in a subtractive mode and act as a prefilter stage for the 0.6 m single-grating spectrograph stage. Holographic gratings blazed for high efficiency at 250 nm are a standard option on commerical spectrometers. The detector is an imaging multichannel photomultiplier tube (IPMT). The IPMT consists of a photo cathode, three microchannel plate arrays, and a resistive anode charge collector. The IPMT quantum efficiency at 250 nm is 15%. The spectral resolution is determined by the spectrograph grating and the spatial resolution of the detector. Using a 2400 groove/mm grating in 2nd order, the resolution of the detector is ca. 4 cm^{-1}. Higher spectral resolution is possible using a CCD detector due to the improved spatial resolution.

UV Raman Spectrometer

Fig. 5.2 Schematic diagram of the ultraviolet Raman spectrometer.

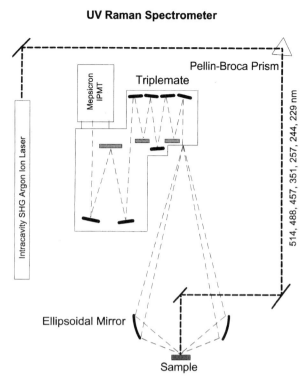

The performance of triple-grating instruments has improved with the availability of instruments using aberation corrected optics that minimize losses due to poor focusing at the slit between the prefilter and spectrograph stages. However, the reflection losses limit the throughput of these instruments to a few percent. Higher throughput instruments operating at visible wavelengths, which make use of a notch filter to attenuate the Rayleigh scattering at the excitation laser wavelength combined with a single-grating spectrograph, have been available for a number of years. Unfortunately, filters for ultraviolet wavelengths with comparable performances are currently not available. We have tested prototype filters with attenuations of 10^{-4} at 244 nm, but these limit the recording of Raman shifts to values above ca. 700 cm^{-1} due to transmission losses near the 244 nm excitation wavelength.

5.2.2
The Fluidized Bed Sample Cell

The advantages of UV Raman spectroscopy in avoiding fluorescence from coke deposits and catalyst impurities and interference from luminescence at elevated temperatures were alluded to above. When it became apparent that the fluorescence problem was avoided, measurements were attempted in the author's laboratory on hydrocarbons adsorbed in catalytic zeolites and during hydrocarbon conversions under catalytic reaction conditions. The spectra suggested that interference from sample damage caused by the ultraviolet laser was a serious problem. One example is cited.

Ultra-stable Y-zeolites with adsorbed benzene, naphthalene, and heptane [17] were examined to evaluate the capabilities of UV Raman spectroscopy for identifying and distinguishing potential coke precursors. The samples were characterized by NMR to ensure that the hydrocarbons remained unreacted following adsorption. The UV Raman spectrum of adsorbed n-heptane appeared to be very similar to that of liquid heptane. In contrast, the spectra from benzene and naphthalene resembled that of coke, with almost no trace of a spectrum representative of the unreacted molecules. In fact, the appearance of a black spot on the sample at the position of the incident laser was clear evidence of sample decomposition. Despite measures to avoid sample heating by reducing the laser power, defocusing the laser beam, and spinning the sample, a spectrum dominated by the formation of coke was still obtained. Attempts to measure Raman spectra using visible and lower energy ultraviolet excitation failed due to fluorescence interference, but the formation of a black spot was not observed. These results strongly suggest that hydrocarbon decomposition was produced by a photochemical rather than a thermal mechanism. Indeed, benzene and naphthalene absorb strongly at the UV excitation wavelengths (257 nm and 244 nm) used for these measurements, whereas *n*-heptane does not.

The effect of photochemically-induced damage on the measured Raman spectrum can be avoided by significantly reducing the build-up of decomposition products in the volume sampled by the Raman spectrometer (typically a cylindrical volume of 100 µm × 100 µm). For liquids, it was found that this could be accomplished by stirring so that decomposition products are continuously flushed from the sampling volume. This observation led to the idea of performing UV Raman measurements on catalytic samples as a fluidized bed (i.e. to make the solid catalysts behave like a "liquid").

Figure 5.3 shows a schematic diagram of the fluidized bed reactor. A stainless steel porous disc (pore size 40 µm) is positioned near the top of a stainless steel tube. The catalyst is placed on top of the porous disc. The tube is surrounded by a cylindrical quartz cover. Gases are introduced into the reactor with the direction of gas flow shown in the diagram. A cylindrical furnace surrounds the reactor. The temperature of the reactor is monitored and controlled by a thermocouple inserted from the bottom and connected to a temperature controller. The reactor is securely attached to a baseplate, which also holds an electromagnetic shaker to facilitate tumbling of the catalyst particles. The laser beam is focused vertically down onto the top surface of the powder bed, and the scattered light is collected by the ellipsoi-

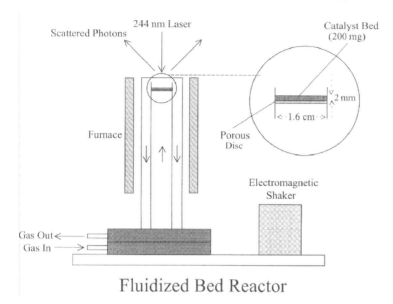

Fig. 5.3 Schematic diagram of the fluidized bed reaction cell for in situ catalytic reaction studies.

dal mirror. The gas flow rate and shaker amplitude are adjusted to produce movement of the catalyst particles (stirring) without lifting the particles out of the bed.

The success of the fluidized bed in eliminating interference from thermal or photochemical decomposition products can be seen in the spectra for benzene adsorbed in H-USY shown in Fig. 5.4. The strong peak at 990 cm^{-1} in liquid benzene (5.4a) is assigned to a symmetric ring breathing mode. This band is barely detectable in spectra 5.4b and 5.4c, which correspond to measurements on a stationary benzene/H-USY pellet and a spinning benzene/H-USY pellet, respectively. Intense coke peaks are present in both of these spectra. Spectrum 5.4d, recorded with the fluidized bed, shows that the dominant species in the spectrum is undecomposed benzene. In fact, even the small peak at 1620 cm^{-1} can be attributed to sample ageing (3+ years) rather than laser-induced decomposition.

Similar tests of the fluidized bed method have been successful using a variety of molecular adsorbates and catalysts (other zeolites, supported oxides, naphthalene, pyridine, methanol, paraffins, olefins, acetonitrile, ammonia, etc.). Some of these results have recently been published [18]. We believe that this fluidized bed method is a major step forward for in situ catalytic reaction measurements using UV Raman spectroscopy. It should also be a useful method for catalytic kinetic measurements by reducing heat and mass transfer effects that arise when catalysts are used in the form of pellets. In the limit of low conversions, the mathematical description of this reactor is the same as that for a short, fixed-bed reactor.

Benzene Raman Spectra

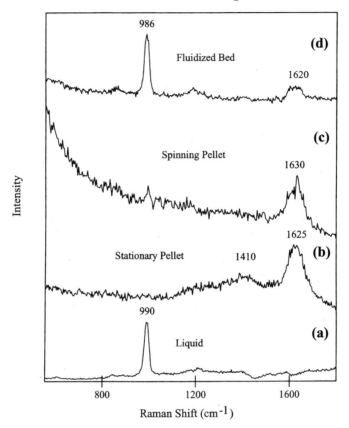

Fig. 5.4 **(a)** Spectrum of liquid benzene; laser power = 5 mW; collection time = 120 s. **(b)** Spectrum of benzene/H-USY recorded on a stationary powder; 2 mw; 900 s. **(c)** Spectrum of benzene/H-USY recorded on a spinning disc; 2 mW; 180 s. **(d)** Spectrum of benzene/H-USY recorded using the fluidized bed apparatus; 2 mW; 3600 s.

However, the fluidized bed should be less susceptible to thermal transients and mass transport effects.

The generality of the fluidized bed approach for avoiding sample decomposition can be evaluated by considering the relationship between the particle motion, the rate of photodegradation, and the fraction of the sample remaining intact. Assuming single-photon photochemistry in the degradation process, the fraction of intact sample is given by:

$$N/N_0 = \exp\left[-\frac{P \cdot \sigma \cdot t}{A}\right] \qquad\qquad (1)$$

where N/N_0 is the fraction of intact sample, P is the laser power in photons per second (~10^{15} for 1 mW at 244 nm), σ is the photochemical cross-section in cm^2 per molecule, t is the residence time of any given area in the laser beam ($<10^{-2}$ s), and A is the area of the focused laser beam (~2×10^{-5} cm^2). An upper limit to σ for organic molecules can be estimated from their absorption cross-sections assuming quantum yields of unity. For benzene, $\sigma = 10^{-19}$ cm^2 and the fraction of intact sample is 1.0. For naphthalene, $\sigma = 10^{-18}$ cm^2 ($\epsilon \approx 1000$) and the fraction of intact sample is 0.97. Even for an extremely photosensitive molecule with $\sigma = 10^{-16}$ cm^2 ($\epsilon = 10^5$), the intact fraction is near 0.5, i.e. measurement of the desired sample spectrum is still feasible.

5.2.3
Raman Tribometer

The second type of in situ Raman spectroscopy measurement carried out in the author's laboratory is designed to monitor the chemistry of a lubricant film in a high-pressure contact between two sliding surfaces. A schematic diagram of the instrument is shown in Fig. 5.5. The excitation laser is focused to ~40 µm at the contact between a steel ball and a rotating sapphire disk (shown end-on). The steel ball (AISI E52100 chrome steel) has a diameter of 38.1 mm and is fixed to provide a pure sliding contact against the window. The sapphire window has a thickness of 1.59 mm and can be rotated at various speeds to provide a shear force to the lubricants. A steel support is used to supplement the mechanical strength of the sapphire window, enabling high pressure to be applied to the contact. The minimum force applied in these experiments was 43.3 N. The calculated contact area was 0.064 mm^2,

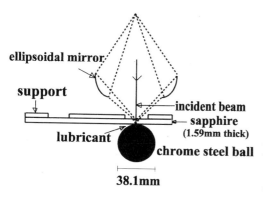

ellipsoidal mirror

support

lubricant

incident beam
sapphire
(1.59mm thick)
chrome steel ball

38.1mm

Fig. 5.5 Schematic diagram of the in situ Raman tribometer. The sapphire window is in the shape of a disk and is shown end-on.

which gives a Hertzian pressure of 0.7 GPa. The maximum force applied was 114.1 N. The calculated contact area was 0.12 mm², which gives a Hertzian pressure of 0.96 GPa. The average sliding speed was about 10 cm s⁻¹. A lubricant was applied to the contact center at the beginning of the experiments. Raman spectra were measured during and after sliding to monitor the lubricant degradation and to identify the nature of the decomposition products.

5.3
Two Examples of Results

5.3.1
Coke Formation in the Methanol-to-Gasoline Reaction

One of the most extensively studied heterogeneous catalytic reactions in recent years has been the conversion of methanol to gasoline-range hydrocarbons by HZSM-5. Indeed, elegant NMR studies on this reaction by Professor James Haw

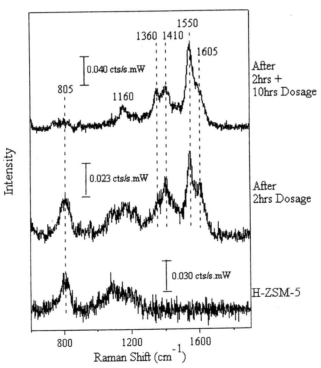

Fig. 5.6 In situ Raman spectra measured following reaction of methanol over H-MFI catalyst at 200°C.

have provided considerable insight into its mechanism [19–22]. As with many hydrocarbon conversion reactions, the activity of methanol-to-gasoline catalysts decreases over time due to the build-up of high molecular weight carbonaceous deposits (coke). Early work in the author's laboratory [23] showed that UV Raman spectra are sensitive to the presence and chemical nature of coke deposits, which prompted a series of UV Raman experiments on coke formation during the methanol-to-gasoline reaction. This work is summarized here. A full account of the research can be found in [24].

Raman spectra were recorded from the HZSM-5 catalyst following a series of exposures to methanol at temperatures of 200°C, 270°C, and 360°C in flowing helium using the fluidized bed reactor. Figures 5.6 and 5.7 show the spectra measured following methanol exposure at 200°C and 270°C, respectively. The attenuation of the broad zeolite bands centered at 800 cm^{-1} and in the region 1000–1300 cm^{-1} arises from strong ultraviolet light absorption by the coke deposit. At low coke loadings, this attenuation can be used as a measure of the coke build-up. By this criterion, the coke loading after a 2-hour exposure to methanol at 200°C is very low.

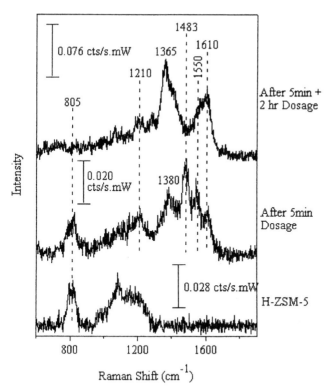

Fig. 5.7 In situ Raman spectra measured following reaction of methanol over HZSM-5 catalyst at 270°C.

The vibrational peak assignments are given in Tab. 5.1. The assignment of all the bands to CC rather than CH vibrations was confirmed by experiments using deuterated methanol. The peak at 1483 cm⁻¹ is characteristic of five-membered-ring aromatic structures, hence the assignment to the cyclopentadienyl symmetric stretch. The band at 1545–1550 cm⁻¹ is assigned to a conjugated olefin having six to eight double bonds [25, 26]. The relative intensities of the bands at ~1610 and ~1400 cm⁻¹ are consistent with a linear, fused ring aromatic structure such as that in anthracene or phenanthrene, rather than a two-dimensional structure such as that in pyrene. Finally, a peak at 2920–2950 cm⁻¹ (not shown) is indicative of aliphatic rather than aromatic CH stretching vibrations, indicating that the coke has an alkylpolyaromatic structure.

A mechanism for aromatic hydrocarbon formation that is consistent with the Raman spectra has been proposed by Schultz and Wei [27], as shown in Scheme 5.1. Methyl-substituted cyclopentenyl carbenium ion is formed by ring-closure following protonation of a conjugated olefin. Deprotonation of the cyclopentenyl carbenium ion leads to the alkyl-substituted cyclopentadienyl species observed by Raman spectroscopy. Hydride removal to form a secondary carbenium ion leads to ring-expansion. Subsequent deprotonation forms the alkyl aromatic structure.

Note that conjugated polyolefin and aromatic coke are formed at 200°C, but with no detectable cyclopentadienyl intermediate. This suggests that another mechanism for coke formation from conjugated olefins is operative at lower temperatures, besides the mechanism shown in Scheme 5.1. One possibility is outlined in Scheme 5.2.

It is interesting to note that the absence of a detectable cyclopentadienyl intermediate at 200°C is coincident with the low loading of hydrocarbon deposit on the catalyst as a result of the low activity in methanol conversion at this temperature. When dimethyl ether is used as the reagent at 200°C, the catalytic conversion and the hydrocarbon loading on the catalyst are much higher, and the cyclopentadienyl intermediate is observed. It is possible that the hydrocarbon deposit acts as a proton reservoir for the proton-exchange reactions required by Scheme 5.1. This view, in which the hydrocarbon deposit acts as a co-catalyst, is consistent with proposals in the literature regarding the formation of catalytically active hydrocarbon deposits during the temperature-dependent induction period observed for the metha-

Raman Shift (cm⁻¹)	Vibrational Peak Assignments
1605–1615	Polyaromatic ring stretch
1360–1410	
1160–1200	Polyaromatic C–C stretch
1545–1550	Conjugated olefin C=C stretch
1483	Cyclopentadienyl symmetric C=C stretch

Tab. 5.1

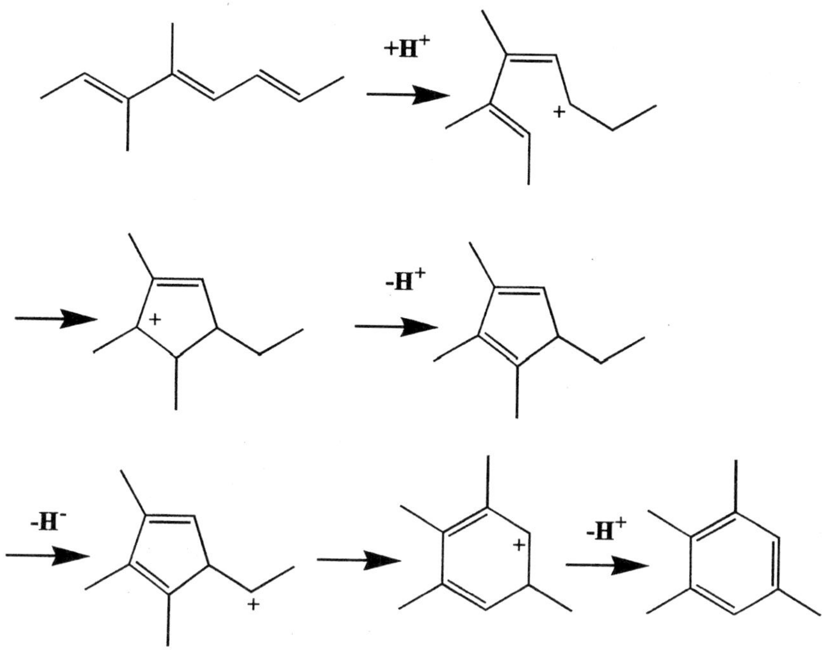

Scheme 5.1

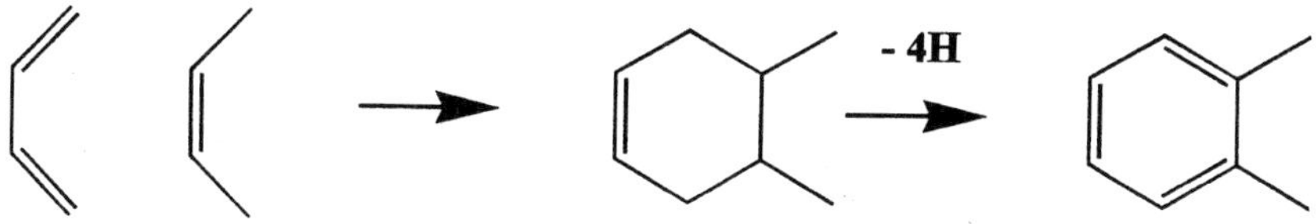

Scheme 5.2

nol-to-gasoline reaction. These proposals have suggested that the hydrocarbon deposits serve as a carbon pool in the reaction [21, 27]. Our results suggest that these deposits may also act as a proton or hydride ion pool.

5.3.2
Lubricant Chemistry

The chemistry of a series of perfluoropolyalkyl ether (PFPE) lubricants during sliding was monitored by in situ ultraviolet Raman spectroscopy using the Raman tribometer. The PFPEs studied are identified by the tradenames Fomblin Z and Krytox. Fomblin Z is a linear polyether containing a mixture of $-OCF_2CF_2O-$ and $-OCF_2O-$ units in an approximately 2:3 ratio. Krytox is a branched polymer made up of $-OCF(CF_3)CF_2O-$ units. These materials are noted for their thermal and oxidative chemical stability [28, 29]. They are used as hard disk lubricants as well as in high temperature and aerospace applications [30–33]. Though PFPEs possess excellent thermal stability in a metal-free environment, the temperature limit of their stability is significantly decreased to about 180 °C when metal alloys or metal oxide surfaces are present [34]. This poses a major problem in the practical utility of these lubricants since metals and metal oxides are prevalent in tribological operations. The goal of the research summarized here was to study in situ the lubricant chemistry of three PFPEs, Krytox 479, Fomblin Z 497, and Fomblin Z 491, using UV Raman spectroscopy in order to better understand the factors that influence their stabilities in a sliding contact. Fomblin Z 497 and Fomblin Z 491 have the same molecular structure but differ in their average molecular weight, 4000 g mol^{-1} and 10,400 g mol^{-1}, respectively, and hence in their viscosity. A full account of this work can be found in [35].

Raman spectra of the lubricants were recorded before, during, and after sliding between the sapphire window and both a chrome steel and a sapphire ball. The balls were not allowed to roll in order to produce a pure sliding contact at 10 cm s^{-1}. The lubricants were applied to the sapphire window with a disposable pipette at the beginning of each experiment and were not replenished during the testing.

The Raman spectra of neat Krytox and Fomblin liquids are shown in Fig. 5.8. Assignments of the Raman shifts, as determined by Pacansky et al. [36], are presented in Tab. 5.2 and Tab. 5.3. Figure 5.9 compares the Raman spectra of Krytox 479 and Fomblin 497 after sliding on the sapphire ball and on the chrome steel ball. The Raman peaks below 800 cm^{-1} and the sharp peak at 1500 cm^{-1} are due to sapphire. The small peak at ~800 cm^{-1} and the structure between 1100 and 1400 cm^{-1} are due to the undecomposed lubricant. The pronounced Raman peaks around 1350 cm^{-1} and 1605 cm^{-1} that are present after sliding on the chrome ball are attributed to the

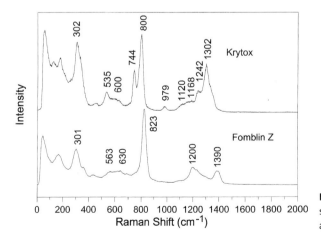

Fig. 5.8 Raman spectra of neat Krytox and Fomblin Z liquids.

Krytox 479

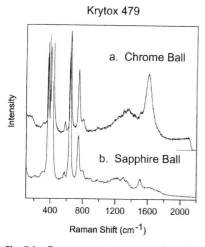

Fomblin Z 497

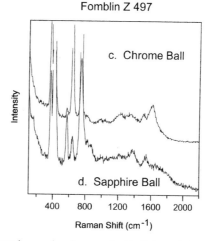

Fig. 5.9 Raman spectra measured in the contact region of the Raman tribometer following sliding between the sapphire window and a chrome steel ball (a and c) and a sapphire ball (b and d).

product of lubricant degradation. The product peaks are also much smaller during sliding, but develop reproducibly after the sliding motion has stopped. Evidence of lubricant degradation is not observable after sliding against the sapphire ball. There is also no evidence of lubricant degradation using Fomblin 491, the higher viscosity Fomblin Z lubricant.

The Raman shifts at 1350 cm^{-1} and 1600 cm^{-1} closely resemble those of amorphous carbon. Amorphous carbon, by its very name, means carbon that exists in an amorphous form, having imperfect graphitic structures such as random orien-

Tab. 5.2 Raman Shifts and Assignments for Krytox

Raman Shifts (cm⁻¹)	*Raman Peak Assignments*
302, 329	Bending modes of the COC skeleton
535	Valence angle bend of CCC chain-rocking and wagging motions of CCO and end CO
600	Stretch and angle bends of end CF_3
744	A combination of pendant CF_3 stretch, angle bend, and CF_2 angle bends
800	COC stretches and CF stretches
979	CC stretch, CF_3 stretch, and CF_2 stretch
1120, 1168	CO stretching
1242	CC stretch of the backbone chain and CF_3 stretch of the pendant group
1302	A combination of CC stretch and pendent CF_3 stretch

Tab. 5.3 Raman Shifts and Assignments for Fomblin

Raman Shifts (cm^{-1})	Raman Peak Assignments
301	Bending mode of COC skeleton
560	Rocking & wagging motions of CCO and CO stretch
630	Stretching and angle bends of end CF_3
823	Combination of COC and CF stretches
1200	CO stretches of CF_2O unit and propylene oxide
1390	CF_2 stretches of CF_2O unit & propylene oxide unit & contribution from CO stretches

tation of the layer planes, random angular displacement of the layers, and irregular overlapping of the layers on top of one another. Raman spectra of various forms of amorphous carbon materials have been reported and analyzed in the literature [37–41]. Two distinct bands characterize the Raman spectra of amorphous carbon: one at around 1350 cm^{-1} and the other at around 1600 cm^{-1}. The relative intensities of these two bands change and their peak positions shift slightly depending on the structure of the material. The peak at around 1600 cm^{-1} is usually referred to as the G band since it is characteristic of single-crystal and polycrystalline graphite [42]. This peak has been assigned by Tuinstra and Koenig [43] as the E_{2g} mode of zone-center phonons in graphite. The peak at around 1350 cm^{-1} is referred to as the D band, after *defects,* since this peak is absent in single-crystal graphite but occurs together with the G band in spectra of polycrystalline graphites, and is assigned to the scattering due to the A_{1g} mode of zone-edge phonons [43]. In a single crystal, this mode is Raman-inactive because the changes in polarizability cancel over the crystal. For small crystallites, this mode is activated by disorder. A comparison of the Raman spectra of the degradation products from Krytox and Fomblin 497 (distinct 1350 cm^{-1} peak; FWHM of the 1600 cm^{-1} band ~80 cm^{-1}) with the as-deposited and annealed amorphous carbon films indicates that the reaction products from the chemical breakdown of the oils are similar in structure to the annealed amorphous carbon film [37, 44].

Our sliding experiments using a sapphire ball were performed to determine whether the amorphous carbon formation showed a material dependence. It is clear from Fig. 5.9 that lubricants that degraded to produce amorphous carbon on the chrome steel ball did not form amorphous carbon on the sapphire ball. The peak at around 1370 cm^{-1} in spectrum 5.9d originates from the sapphire ball. The chemical stability of Krytox and Fomblin on the sapphire ball demonstrates the crucial role ball surfaces play in the chemistry of PFPE fluids. These results confirm the argument that surface activity is an important factor with regard to the chemical degradation of PFPE fluids. The stability of Krytox and Fomblin on the sapphire ball may stem from the fact that aluminum oxide is, in general, relatively inert chemically. At room temperature, aluminum oxide incorporates weakly adsorbed OH groups and water molecules. Water adsorbed on the aluminum atom forms rather weak Brønsted acidic sites, and this acidity is not sufficiently strong to catalyze the degradation of the Krytox and Fomblin lubricants [45].

A number of factors may explain why Krytox 479 and Fomblin 497 formed amorphous carbon while Fomblin 491 did not. These include differences in flash temperatures from asperity contacts due to film failure, the larger fraction of carbon in the form of CF_3 in Krytox and Fomblin 497, and the presence of impurities. However, with our current understanding of this chemistry, it is not apparent what is responsible for the absence of amorphous carbon formation with Fomblin 491.

5.4
Summary

Ultraviolet Raman spectroscopy is a powerful new tool for the analysis of technologically important materials and for materials chemistry in general. The fact that fluorescence interference is practically never a problem means that a Raman spectrum can be measured without regard for the purity of the materials or the chemical reaction conditions. At the present time, there are four research groups around the world (Dalian Institute, Eindhoven, Worcester Polytechnic, and UC Berkeley) with active programs using UV Raman spectroscopy. This list will surely grow with improvements in instrumentation and as the problems associated with ultraviolet photodegradation and time resolution are conquered.

Acknowledgements

Financial support of this work was provided by the Department of Energy, Office of Basic Energy Sciences, Division of Chemical Sciences, under contract no. DE-FG02-97ER14789, by the Northwestern University Center for Catalysis and Surface Science, and by the Center for Surface Engineering and Tribology.

References

1 Pittman, R. M.; Bell, A. T., *Catal. Lett. 1994*, **24**, 1–13.
2 Lunsford, J. H.; Yang, X.; Haller, K.; Laane, J.; Mestl, G.; Knoezinger, H., *J. Phys. Chem. 1993*, **97**, 13810–13813.
3 Wachs, I. E.; Hardcastle, F. D., *Proc. 9th Int. Congr. Catal. 1988*, **3**, 1449–1456.
4 Wachs, I. E.; Segawa, K., *Charact. Catal. Mater. 1992*, 69-88.
5 Dixit, L.; Gerrard, D. L.; Bowley, H. J., *Appl. Spectrosc. Rev. 1986*, **22**, 189–249.
6 Wachs, I. E.; Hardcastle, F. D., *Catalysis 1993*, **10**, 102–153.
7 Knoezinger, H.; Mestl, G., *Top. Catal. 1999*, **8**, 45–55.

8 Mestl, G., *J. Mol. Catal. A: Chem.* *2000*, **158**, 45–65.

9 Stencel, J. M., *Raman Spectroscopy for Catalysis*, Van Nostrand Reinhold: New York, 1990.

10 For correlations between metal-oxygen stretching frequencies and bond lengths see, for example: Hardcastle, F. D.; Wachs, I. E., *J. Phys. Chem.* *1991*, **95**, 10763, and references therein.

11 Asher, S. A.; Johnson, C. R., *Science (Washington DC 1883–) 1984*, **225**, 311–313.

12 Ferraro, J. R.; Nakamoto, K., *Introductory Raman Spectroscopy;* Academic Press: Boston, 1994.

13 Asher, S. A., *Anal. Chem. 1993*, **65**, 201.

14 Stair, P. C.; Li, C., *J. Vac. Sci. Technol., A 1997*, **15**, 1679–1684.

15 Li, C.; Stair, P. C., *Stud. Surf. Sci. Catal. 1996*, **101**, 881–890.

16 Cheong, C. U.; Stair, P. C., *Book of Abstracts, 213th ACS National Meeting, San Francisco, April 13–17, 1997*, COLL-085.

17 Courtesy of Dr. Jeffrey Miller, BP Amoco Chemicals.

18 Chua, Y. T.; Stair, P. C., *J. Catal. 2000*, **196**, 66–72.

19 Munson, E. J.; Kheir, A. A.; Lazo, N. D.; Haw, J. F., *J. Phys. Chem. 1992*, **96**, 7740–7746.

20 Munson, E. J.; Lazo, N. D.; Moellenhoff, M. E.; Haw, J. F., *J. Am. Chem. Soc. 1991*, **113**, 2783–2784.

21 Goguen, P. W.; Xu, T.; Barich, D. H.; Skloss, T. W.; Song, W.; Wang, Z.; Nicholas, J. B.; Haw, J. F., *J. Am. Chem. Soc. 1998*, **120**, 2650–2651.

22 Haw, J. F.; Nicholas, J. B.; Song, W.; Deng, F.; Wang, Z.; Xu, T.; Heneghan, C. S., *J. Am. Chem. Soc. 2000*, **122**, 4763–4775.

23 Li, C.; Stair, P. C., *Stud. Surf. Sci. Catal. 1997*, **105A**, 599–606.

24 Chua, Y. T.; Stair, P. C., to be published.

25 Ivanova, T. M.; Yanovskaya, L. A.; Shorygin, P. P., *Optics and Spectroscopy 1965*, **18**, 115–118.

26 Baruya, A.; Gerrard, D. L.; Maddams, W. F., *Macromolecules 1983*, **16**, 578–580.

27 Schulz, H.; Wei, M., *Micro. Meso. Mater. 1999*, **29**, 205.

28 Sianesi, D.; Zamboni, V.; Fontanelli, R.; Binaghi, M., *Wear 1971*, **18**, 85–100.

29 Helmick, L. S.; Jones, R. W., *Lubr. Eng. 1994*, **50**, 449–457.

30 Carre, K. J., *ASLE Trans. 1986*, **29**, 121–125.

31 Gumprecht, W. H., *ASLE Trans. 1966*, **9**, 24–30.

32 Snyder, C. E.; Dolle, R. E., *ASLE Trans. 1976*, **19**, 171–180.

33 Zaretsky, E. B., *Trib. Int. 1990*, **23**, 75–93.

34 Koka, R.; Armatis, F., *Trib. Trans. 1997*, **40**, 63–68.

35 Cheong, A.; Stair, P. C., *Trib. Lett. 2001*, **9**.

36 Pacansky, J.; Miller, M.; Hatton, W.; Liu, B.; Scheiner, A., *J. Am. Chem. Soc. 1991*, **113**, 329–343.

37 Bowden, M.; Gardiner, D. J.; Southall, J. M., *J. Appl. Phys. 1992*, **71**, 521.

38 Cuesta, A.; Dhamelincourt, P.; Laureyns, J.; Martinez-Alonso, A.; Tascon, J. M. D., *Carbon 1994*, **32**, 1523.

39 Dillon, R. O.; Woollam, J. A., *Phys. Rev. B 1984*, **29**, 3482–3489.

40 Espinat, D.; Dexpert, H.; Freund, E.; Martino, G.; Couzi, M.; Lespade, P.; Cruege, F., *Appl. Catal. 1985*, **16**, 343–354.

41 Tamor, M. A.; Vassell, W. C., *J. Appl. Phys. 1994*, **76**, 3823–3830.

42 Schwan, J.; Ulrich, S.; Batori, V.; Ehrhardt, H., *J. Appl. Phys. 1996*, **80**, 440–447.

43 Tuinstra, F.; Koenig, J. L., *J. Chem. Phys. 1970*, **53**, 1126–1130.

44 Wagner, J.; Ramsteiner, M.; Wild, C.; Koidl, P., *Phys. Rev. B 1989*, **40**, 1817–1823.

45 Tanabe, K., *Solid Acids and Bases;* Academic Press: New York, 1970.

6
In Situ Infrared Methods

Russell F. Howe

6.1
Introduction

Infrared spectroscopy has a long history as a technique for studying catalysts and adsorbed molecules. Early Russian work by the group of Terenin as early as 1940 [1] was not readily accessible at the time, and the pioneering work of Eischens [2], Sheppard [3], and Peri [4] is credited with bringing the method to the wider attention of the catalyst community by the early 1960s. Subsequently, numerous monographs and review articles attest to the importance of infrared spectroscopy in modern catalyst research. The recent review by Busca [5], specifically on metal oxide catalysts, traces some of the historical background and compares infrared spectroscopy with other methods for obtaining vibrational spectra (Raman, EELS, IETS, INS). This review should be consulted for references to earlier reviews and monographs. Infrared spectroscopy is also comprehensively covered in a recent monograph on spectroscopic methods in catalysis [6].

In situ infrared spectroscopy can be defined as the observation of catalyst and adsorbed species at reaction temperature in the presence of reactants. The advent of fast scanning Fourier transform instruments in the past 25 years has greatly eased the experimental difficulties of in situ measurements. Spectra can now be collected in short times with good signal-to-noise from poorly transmitting samples. Infrared emission from hot catalyst samples is less of a problem than it was with wavelength scanning instruments, and subtraction of contributions from gas-phase species is easily managed. Alternative sampling methods, such as diffuse reflectance, microspectroscopy on zeolite single crystals, reflection absorption from single crystal surfaces, and infrared emission measurements have all become possible with the increased sensitivity, as discussed further below. Other new developments include the use of picosecond infrared laser spectroscopy to probe the dynamics of adsorbate-catalyst interactions, and the introduction of infrared-visible sum-frequency generation spectroscopy as a surface-specific technique for obtaining infrared spectra of species adsorbed on single crystal surfaces at high gas pressures.

In this chapter, selected developments in in situ infrared spectroscopy of catalysts over the past five years are reviewed. No attempt is made at complete literature coverage; instead, selected examples of the different types of experiments currently being reported are presented. The intent is to illustrate the breadth and potential opportunities of the field, rather than to catalogue the recent literature. A recent review by Lercher et al. [7] has nicely outlined the strategies that are employed in in situ infrared studies, and the contributions that such studies can make to catalyst and process development.

6.2
Experimental Aspects

This section deals with some recent developments in cell designs and experimental techniques for in situ infrared studies.

6.2.1
In situ cells for transmission spectroscopy

Transmission measurements through pressed disks of catalyst have been the standard method employed in infrared studies for more than 50 years. Over this period, numerous designs of in situ cell have been proposed and implemented. In all cases, the requirements are to hold the catalyst disk securely at a controlled temperature while it is exposed to reactant gases. In order to relate the observed spectra of adsorbed species to catalytic reactions, the cell should be configured to act as a microreactor. This can be achieved in pulse mode, or in continuous flow stirred tank reactor mode. An example of the latter has been described by Mirth et al. [8] and is shown schematically in Fig. 6.1. Features of this design are a sample holder containing a heating element suspended from the lid of the cell, and a cell body containing CaF_2 windows sealed with viton o-rings, a gas inlet, and a gas outlet. The cell is stated to operate over a temperature range from 300 to 870 K and up to 5 bar total pressure. The minimal void volume between the cell windows of 1.5 cm^3 serves two purposes: the pathlength of the infrared beam through the gases in the cell is minimized, reducing the contribution to spectra from gas-phase bands, and the mixing of gases in the cell is effective. A transient response time of less than 20 seconds was obtained at a total flow rate of 20 mL per minute. Such a reactor is connected to the usual reactant and product gas lines, and catalyst per-

PART (a)

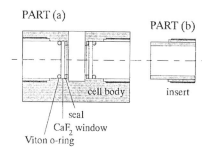

PART (b)

insert

cell body

seal

CaF₂ window

Viton o-ring

Fig. 6.1 Schematic of in situ infrared cell described by Mirth et al. [8]. Reproduced with permission.

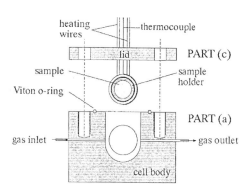

heating wires

thermocouple

lid

PART (c)

sample

sample holder

Viton o-ring

gas inlet

PART (a)

gas outlet

cell body

formance is monitored by gas chromatographic or mass spectrometric analysis. An example of data collected from this reactor is described in Section 6.3.1.3

In some situations, it is desirable to record spectra below room temperature, either to examine the spectra of physically adsorbed probe molecules or to study reaction pathways in low temperature reactions. Cells for this purpose must incorporate both heating elements and liquid nitrogen cooling lines. An elegant example of such a cell capable of operating between 120 K and 773 K has been described by Mariscal et al. [9].

An infrared cell operating in pulse reactor mode must be capable of rapidly responding to the introduction of reactant pulses. This is achieved by reducing the void volume around the sample even further. The use of transient in situ infrared methods by the group of Chuang has relied on a cell design in which the cylindrical cell body is heated externally, and the sample disk is sandwiched between two infrared transmitting CaF_2 rods, which reduce the void volume within the cell to 0.05 cm³ [10]. This cell is stated to operate up to 773 K and 60 bar pressure, and is connected to an appropriate pulse injection system inlet with effluent analysis by gas chromatography and mass spectrometry. This particular cell has the advantages of ease of assembly and a particularly low void volume. Small void volumes and short gas-phase pathlengths can also be obtained by sandwiching the sample disk between two flanges fitted with windows and plate heaters, but in this case the upper temperature obtainable is limited by the window seals to ca. 573 K [11].

In situations where the response time is not a critical factor, in situ cells can be of a much simpler design. In the author's laboratory, a modified version of a design first published by Moon et al. [12], in which the sample disk is placed in an externally heated quartz tube terminated by water-cooled stainless steel window mountings, has been used as a pulse reactor at temperatures up to 1000 K (although infrared emission from cell and sample prevent useful spectra being recorded much above 800 K). An interesting low-budget mini-cell assembled almost entirely from commercial swagelock components, reported by Komiyama and Obi [13], uses a sheathed thermocouple as a micro-heater with an extremely rapid temperature response.

The final choice of cell design for an in situ transmission experiment depends on the nature of the reaction being studied and whether continuous flow or pulse methods are being employed. The many ingenious designs now available in the literature have largely overcome the apparent incompatibilities between high-temperature materials, infrared transmitting windows, and short pathlength minimal void space requirements.

6.2.2
Diffuse Reflectance Spectroscopy

The improved signal-to-noise available with FTIR instruments fitted with liquid nitrogen cooled mercury cadmium telluride (MCT) detectors has stimulated increased use of diffuse reflectance as an alternative to transmission for obtaining spectra of catalysts and adsorbed molecules. In principle, the diffuse reflectance (DRIFT) technique offers several advantages over transmission. Sample preparation is more straightforward: powdered catalysts can be examined without the need to press them into disks. Some catalyst materials (depending on particle size and morphology) can be difficult or impossible to press into disks that are both thin enough to transmit infrared radiation and robust enough to survive in situ treatment. Diffuse reflectance measures the scattered radiation, which is complementary to that measured in transmission; i.e. those samples that give poor transmission spectra because of scattering losses should give good DRIFT spectra, and vice versa. Powdered catalyst samples may also be less susceptible to diffusion limitations than the corresponding pressed disks. On the other hand, cell design for in situ measurements of catalysts by diffuse reflectance is more difficult.

Most DRIFT studies of catalysts carried out to date have used commercially available in situ cells. The catalyst powder is placed in a horizontal sample cup on top of a heated stage, which also contains gas inlet and outlet lines. A shroud placed on top of the sample cup contains infrared-transparent windows, through which the incident infrared beam and diffusely reflected beam can pass. Drochner et al. [14] have pointed out a number of drawbacks with such designs. The pathlength through the gas phase is relatively long, so that achieving satisfactory cancellation of gas-phase bands from the spectra can be difficult. A second cancellation problem is caused by the inevitable deposition of reaction by-products on the cooled windows of the cell. Since the composition of such a deposited layer will be continuously changing, subtraction of the spectrum is difficult. A third difficulty encountered in the author's laboratory is the temperature gradient existing in such cells, particularly at higher temperatures. Optical alignment can also be tedious.

A new cell design with improved performance for in situ DRIFT measurements of catalysts has very recently been described by Drochner et al. [14]. The main new feature of this design is the provision of two sample cups with a sealed mechanism for rapid interchange of the two cups into or out of the incident infrared beam without opening the cell. The catalyst sample is placed in one cup, and a reference material that does not adsorb any of the molecules of interest is placed in the other. By rotating the reference sample into the beam, a background spectrum can be recorded immediately before or immediately after the sample spectrum, allowing, in principle, complete subtraction of gas-phase and window contaminant bands. The other important feature of this design is the use of fritted glass sample cups, which allow uniform flow of reactant gases through the catalyst bed. Drochner et al. published spectra measured using this cell at 573 K, and state that it will operate up to 673 K [14].

Diffuse reflectance can clearly be made to work as an in situ method, although transmission will still be preferred for those catalyst samples which lend themselves to the pressing of thin infrared-transparent disks without a loss of porosity.

6.2.3
Infrared Emission Spectroscopy

According to Kirchoff's law, the infrared emission spectrum of a heated sample contains the same information as the absorption spectrum. The infrared emissivity is proportional to the fourth power of the temperature difference between the emitting sample and the detector, and at medium to high temperatures (below 1000 K) black-body emission shows a maximum in the mid-infrared region of the spectrum. The advantages and disadvantages of measuring emission spectra compared with transmission and diffuse reflectance spectra have been summarized by Sullivan et al. [15]. At temperatures above 573 K, the entire mid-infrared frequency range is, in principle, accessible. Samples can be examined in the form of powders, although it is necessary to keep the powder layer thin to avoid self-absorption effects. However, as pointed out by Sullivan et al., this can be an advantage of emission over transmission. In transmission spectra of oxide or zeolite disks, the characteristic metal-oxygen stretching modes and zeolite lattice bands are normally too intense to be observed. In emission spectra, it is possible to monitor both structural changes in the catalyst and adsorbed species at the same time. Disadvantages of the emission technique, as listed by Sullivan et al., are the interferences from background radiation (particularly if working with a liquid nitrogen cooled detector) and the fact that the signal-to-noise and frequency range accessible are strongly temperature-dependent.

Figure 6.2 shows the schematic design of an infrared emission cell for in situ catalyst studies reported by Sullivan et al. [15]. A thin layer of catalyst powder is deposited on a gold-coated sample stage, which is heated from below by cartridge heaters. Infrared emission from the sample is focussed into the optical bench of an FTIR spectrometer in place of the normal beam from the infrared source (which is turned off). Reactant gases flowing into the cell are mixed by an impeller, and pumped out of the cell to a mass spectrometer for analysis. Preliminary data shown in [15] indicate that the quality of emission spectra obtained from this cell for CO adsorbed on supported metal catalysts and from HZSM-5 zeolite are comparable with the corresponding transmission and diffuse reflectance data, respectively.

An alternative cell design, in which the heated sample is mounted vertically, has been described more recently by Weber et al. [16]. In this design, steps are taken to water-cool the cell body in order to minimize emission from the cell. Attention is also drawn to a miniaturized infrared emission cell described by Vasallo et al. [17], in which the emission source is constructed from an atomic absorption graphite rod furnace and emitted radiation collected from a 3 mm diameter area of a platinum sample cup.

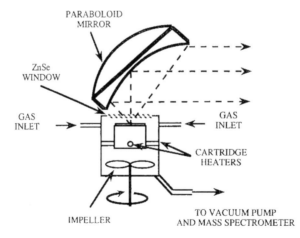

Fig. 6.2 Schematic of in situ infrared emission cell described by Sullivan et al. [15]. Reproduced with permission.

To date, the infrared emission technique has been under-utilized for in situ studies of catalysts. It has, however, exciting potential for the study of high-temperature catalytic reactions, where signal-to-noise will be optimized.

6.2.4
Infrared Microspectroscopy

With a modern infrared microscope, it is possible to obtain good quality transmission infrared spectra from sample areas as small as 20 microns by 20 microns (and this size limit can be further lowered if a synchrotron is used as the source of infrared radiation [18]). Several groups have exploited this technique to study molecules adsorbed in single crystals of zeolite catalysts. Schüth et al. [19] and Lercher [20] have described in situ cells that can be mounted on the sample stage of an infrared microscope, allowing zeolite single crystals to be studied at temperatures up to 400 °C in the presence of adsorbing gases or in vacuo. Cells of this type have been used to study orientations of adsorbed molecules [21] and of zeolite templates, as well as the decomposition of templates during zeolite activation [22, 23].

Of most relevance to catalysis, however, is the use of microspectroscopy to study the diffusion of reactants into the pores of zeolite single crystals. In crystals as large as those needed for single-crystal infrared studies (typically 100 microns), diffusion paths are long and diffusion is inevitably the rate-limiting step in any catalytic reaction. Muller et al. have employed an infrared microscope to study the diffusion of toluene in ZSM-5, by measuring the spectrum as a function of time at different points within the crystal [24]. Studies of this kind represent a powerful application of the infrared microscopy technique, which can be applied in situ to any zeolite catalyst that can be grown in single-crystal form.

A novel alternative experimental approach to spatially resolved infrared spectroscopy of catalysts was reported by Qin and Wolf [25]. These authors used an infrared camera, coupled with a narrow band pass filter set to the frequency of linearly bonded carbon monoxide adsorbed on an Rh-silica catalyst (2160 cm^{-1}), to monitor the spatial distribution of this species across a disk of the catalyst mounted in a conventional transmission in situ cell during CO oxidation. By turning off the infrared source and removing the filter, the infrared camera could also be used to image the temperature distribution. The authors point out the potential of this approach for understanding oscillatory catalytic reactions such as CO oxidation on supported metal catalysts under the reaction conditions used.

6.2.5
Reflection-Absorption Infrared Spectroscopy (RAIRS)

The RAIRS technique is well-developed in surface science as an approach for measuring infrared spectra of adsorbed species on single-crystal metal surfaces [26].

The method involves specular reflection of an infrared beam at close to grazing incidence from a flat surface. Adsorbed molecules present on the surface will absorb part of the infrared beam; this absorption is greatly enhanced if the infrared radiation is polarized perpendicular to the surface, and on metal surfaces the so-called surface selection rule means that only adsorbate vibrations with a dipole moment change perpendicular to the surface will be detected.

The advantage of RAIRS over most other surface science techniques is that spectra can be measured in the presence of gas-phase reactants (particularly if polarization modulation is used to remove gas-phase contributions to the spectra). This advantage is also shared by the much newer method of sum frequency generation spectroscopy described below. In catalysis, RAIRS is restricted to fundamental studies on model systems, but it is making important contributions to the understanding of reaction pathways on metal surfaces (see Section 6.3.3 below). A recent experimental approach to model studies on oxide surfaces is to deposit a thin film of oxide onto a single-crystal metal substrate, and the RAIRS technique is finding wide application to such model oxide catalysts [27].

6.2.6
Sum Frequency Generation Spectroscopy (SFG)

A second, much newer technique for obtaining infrared spectra of well-defined model catalysts under high-pressure, high-temperature conditions is sum frequency generation spectroscopy. SFG is a second-order nonlinear optical process, in which infrared and visible light are mixed to produce radiation with a frequency equal to the sum of the infrared and visible frequencies. An SFG experiment is conducted by holding the visible light component fixed at a single frequency and scanning the infrared frequency through the spectral region of interest. A spectrum is obtained by plotting the SFG intensity versus the frequency of the infrared radiation. An SFG response will be produced whenever the infrared frequency matches a vibrational frequency in the sample that is SFG-active. Since SFG can only occur in environments without inversion symmetry (in the electric dipole approximation), the SFG spectrum of a model catalyst will be dominated by the solid–gas interface, with very little contribution from the underlying solid and none at all from the gas phase [28].

Cremer et al. have reviewed some of the first applications of SFG to in situ studies of model catalysts [29], while the design of an SFG-compatible UHV-high pressure in situ reaction cell has been described by Rupprechter et al. [30].

The complexity of the SFG experiment, requiring a high power pulsed laser, frequency doubling to generate the visible component, nonlinear optical crystals to generate the tunable infrared component, and detection equipment able to discriminate the sum-frequency signal, has meant that few laboratories have so far investigated this technique for in situ catalyst studies. The one major advantage that SFG offers over RAIRS for in situ studies of model catalyst systems is that it can be applied to observe transient species, as recently demonstrated by Domen et al. [31].

6.2.7
Picosecond Infrared Spectroscopy

Picosecond infrared spectroscopy was first developed in the 1970s, and has been widely applied to liquids [32] and in surface science [33]. The technique uses a very short (25 ps) infrared laser pulse to selectively excite one particular vibrational transition. The system's response to the perturbation and the subsequent decay back to equilibrium are monitored with a second time-delayed laser pulse. In terms of catalyst surfaces, three types of information are obtained. The vibrational energy relaxation at a catalytic site (such as a Brønsted acid hydroxyl group) can be observed in real time. The broadening mechanisms behind infrared bands, which may be related to the dynamic interactions of the site involved with reactant molecules, can be investigated. Thirdly, and arguably most importantly from the viewpoint of catalysis, the time scale of the technique allows for the first time direct observation of transition states and reaction intermediates at an active site.

In situ picosecond infrared spectroscopic measurements on catalyst systems have so far been undertaken largely by two groups, the Hirose–Domen group in Japan, and the van Santen–Kleyn–Bakker group in the Netherlands. These two groups have recently published a major review of the technique and the results obtained with it [34].

The major experimental requirement for such spectroscopy is the ability to generate very short and very intense infrared pulses with continuously tunable frequency. With appropriate arrangement of the optical components, two independently tunable pulse trains can be generated from the same laser. Figure 6.3 shows the pump-probe configuration used to either measure the relaxation of an excited vibrational state (where the probe frequency is equal to the pump frequency, but is delayed by a variable amount), or to measure the transient infrared spectra of transition states or reaction intermediates (where the probe frequency is scanned). Spectra are measured by transmission through pressed disks of catalyst in an appropriate in situ cell.

An example of the use of this technique to study the interaction of methanol with hydroxyl groups in zeolites is described in Section 6.3.1.2 below. The current limitations of the method, as summarized by Bonn et al. [34], are that relatively strongly absorbing bands are needed to give reasonable signal-to-noise, the spectral resolu-

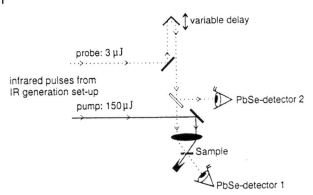

Fig. 6.3 Pump-probe configuration for picosecond infrared spectroscopy (reproduced with permission from [34]). A reference detector (detector 2) is used to correct for shot to shot intensity fluctuations.

tion of the method is limited by the band width of the infrared laser pulses to about 10 cm^{-1}, and the time resolution currently available is insufficient to study some particular processes such as the lifetimes of transition states. For this purpose, high repetition rate femtosecond laser systems now becoming available may provide some advantage, although they suffer from decreased spectral resolution.

6.3
Recent Applications of in situ Infrared Spectroscopy

In this section, selected examples of recent applications of in situ infrared spectroscopy in studying catalytic reactions are discussed. The reader is referred to the original literature for more complete information about each example.

6.3.1
Zeolite Catalysts

The high internal surface area of zeolite catalysts, the comparative ease of preparing pressed disks of zeolites for transmission measurements, and the importance of zeolite catalysts in many industrial processes, have all contributed to the large number of in situ infrared studies of these catalysts. In this section, four different examples are reviewed to illustrate the power of the technique.

6.3.1.1 Low-temperature bond migration in olefins

Kondo and co-workers have carried out a number of low-temperature in situ infrared studies on the interactions of olefins with different acid zeolites [35–40]. In these systems, adsorption and reaction below room temperature were studied in a simple liquid nitrogen cooled low-temperature cell. Double-bond migration in olefins (e.g. 1-butene isomerizing to cis- and/or trans-2-butene) is well-known to be readily catalyzed by Brønsted acid sites, and is normally considered to proceed via a protonated carbenium ion intermediate:

$$CHR{=}CHCH_2R + H^+ \leftrightarrow CH_2R\text{-}CH^+CH_2R' \leftrightarrow CH_2R\text{-}CH{=}CHR' + H^+$$

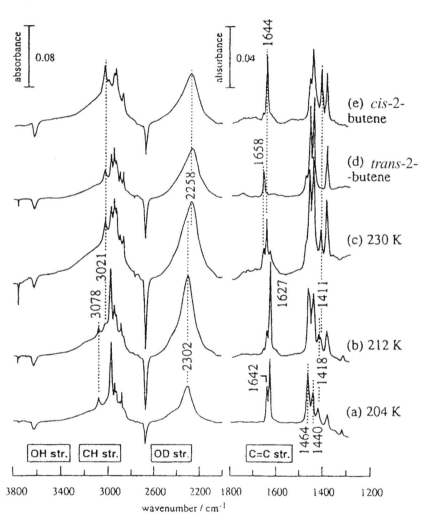

Fig. 6.4 Infrared spectra of 1-butene adsorbed in DZSM-5 at various temperatures. Also shown are reference spectra of *trans*- and *cis*-2-butene adsorbed in DZSM-5. Reproduced with permission from [35].

Kondo et al. examined the interaction of 1-butene with a deuterated ZSM-5 catalyst at temperatures between 204 K and 230 K, and concluded that in this temperature range isomerization occurred without proton transfer [35]. Figure 6.4 shows the spectra obtained.

The spectrum of the DZSM-5 prior to adsorption was in all cases subtracted from the spectra after adsorption. Thus, the spectrum of 1-butene adsorbed at 204 K is assigned to the 1-butene molecule hydrogen-bonded to acidic OD groups; the v(OD) band at 2671 cm^{-1} is shifted to 2302 cm^{-1}, and the adsorbed 1-butene bands are slightly perturbed relative to the free molecule (note, in particular, the C=C stretching mode at 1627 cm^{-1} and the olefinic C–H stretch at 3078 cm^{-1}). On raising the temperature above 212 K the spectrum changes: the perturbed v(OD) band is further shifted to 2258 cm^{-1} and the adsorbed olefin bands now resemble those of a mixture of *cis*- and *trans*-2-butene, an assignment confirmed by adsorbing these molecules separately.

Figure 6.5 shows plots of the integrated intensities of the various bands as a function of temperature. It is particularly noteworthy that over the temperature range in which isomerization of adsorbed 1-butene occurs, there is no change in the relative concentrations of OD and OH groups in the catalyst.

Kondo et al. concluded from these data that the hydrogen-bonded alkenes undergo double-bond migration without proton transfer, since any protonation of the alkene would inevitably lead to OD/OH exchange. On raising the temperature further, exchange does occur; the intensity of the OD band falls and that of the corre-

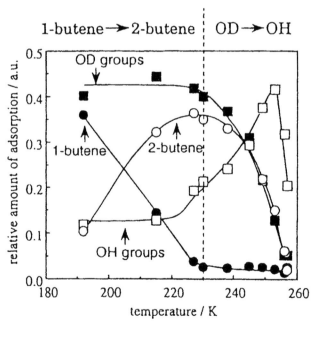

Fig. 6.5 Temperature dependence of the relative concentrations of adsorbed species during reaction of 1-butene over DZSM-5. Reproduced with permission from [35].

sponding OH band rises. At the same time, further reaction of the 2-butene occurs, forming dimeric C_8 products.

Similar low-temperature isomerization without H/D exchange has also been shown to occur on H-ferrierite zeolite [41], showing that the Brønsted acid sites in zeolites behave in a different manner from protic acids in solution. The strong hydrogen bonding of alkenes to the Brønsted sites has been shown to be quite general for many different olefins [38].

6.3.1.2 Methanol conversion over acid zeolites

The interaction of methanol with acid zeolites has received a great deal of experimental and theoretical attention following the discovery and eventual commercialization of the so-called MTG (methanol to gasoline) process [42, 43].

The reaction pathways by which methanol is first dehydrated to dimethyl ether, and then to light olefins, which subsequently convert to aromatics, have been widely investigated spectroscopically, through conventional catalytic measurements of product distributions and reaction kinetics, by the use of isotopic tracers, and theoretically. In situ infrared methods have made important contributions to the understanding of these reaction pathways.

Early in situ studies of the interaction of methanol with the HZSM-5 zeolite catalyst used in the MTG process employed an in situ cell functioning as a pulsed microreactor [44]. Injection of single pulses of methanol at low temperatures (e.g. 423 K) gave spectra that were attributed at the time to protonated methanol species [44, 45]. This assignment was based on loss of the ν(OH) band associated with the Brønsted acid sites in the zeolite, and the appearance of broad, intense ν(OH) bands at lower frequencies. Figure 6.6 shows typical spectra from a more recent publication, in which a series of ZSM-5 zeolite catalysts with differing Si:Al ratios were studied [46]. The difference spectra recorded after injection of a single pulse of methanol in a nitrogen stream at 423 K clearly show the loss of Brønsted acid protons (the negative feature at 3610 cm^{-1}) and the appearance of new lower frequency ν(OH) and ν(CH) bands due to adsorbed methanol. This is most evident in the case of the zeolite containing the highest concentration of Brønsted acid sites (Fig. 6.6(a)). There is also evidence for interaction of methanol with silanol groups on the external surface of the zeolite (the negative band at 3740 cm^{-1}) and in the case of the ZSM-5 zeolite having an Si:Al ratio of 16, with extra-framework AlOH sites (negative band at 3650 cm^{-1} in Fig. 6.6(c)).

The original assignments of the spectrum obtained on injection of methanol into HZSM-5 at 423 K to protonated methanol have now been shown to be wrong. High-level density functional calculations have shown that protonated methanol in ZSM-5 is a transition state rather than a stable species [47, 48], and the pattern of

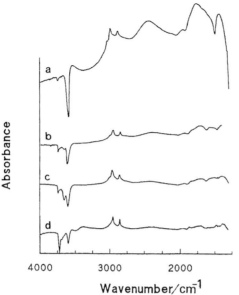

Fig. 6.6 Difference infrared spectra recorded following injection of methanol at 423 K into HZSM-5 zeolites with different Si:Al ratios: (a) 12, (b) 27, (c) 16, (d) 121. Reproduced with permission from [46].

three intense and broad ν(OH) bands has been recognized as being characteristic of strongly hydrogen-bonded methanol [49].

Bonn et al. [50] subsequently applied time-resolved infrared spectroscopy to further investigate the initial interaction of methanol with Brønsted acid sites in a zeolite. The lower sensitivity of the time-resolved technique made it necessary for them to use zeolite HY with a higher concentration of acidic sites than HZSM-5, which leaves open the question as to whether the chemistry observed will be identical to that occurring in HZSM-5.

The experiment performed by Bonn et al. was to selectively excite the adsorbed methanol at a frequency of 3250 cm^{-1} with a 20 ps infrared pulse, and to observe the subsequent changes in the absorption spectrum with a second weak but tunable probe pulse, following a variable delay. The frequency of 3250 cm^{-1} was chosen to correspond to the ν(OH) frequency of hydrogen-bonded methanol.

Figure 6.7 shows the absorption difference spectrum measured by scanning the frequency of the probe pulse when the probe pulse was applied 100 ps before the pump pulse and 250 ps after the pump pulse. In the first case, the absorption spectrum is unchanged and the difference spectrum is a straight line. In the second case, there is a decrease in absorbance at 3250 cm^{-1} and an increase in absorbance at 3580 cm^{-1}.

Bonn et al. attribute these changes to laser-induced dissociation of the hydrogen bond between the OH groups of two adjacent methanol molecules (responsible for the 3250 cm^{-1} infrared band). They argue that such dissociation allows reorienta-

0.06
0.04
0.02
0.00
-0.02
-0.04
-0.06
ln(T/T0)
pump
τdelay= -100 ps
τdelay= 270 ps
calculation
3100
3200
3300
3400
3500
3600
3700
frequency (cm-1)

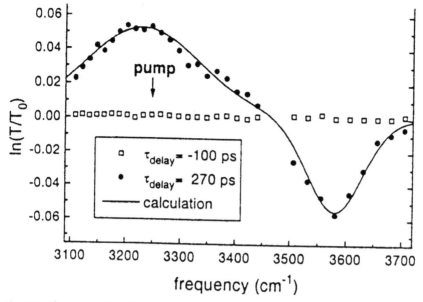

Fig. 6.7 Changes in the infrared spectrum of methanol adsorbed in HY following excitation at 3250 cm^{-1} with a 20 ps pulse. Reproduced with permission from [34].

(a)
H3C
H
~3250 cm-1
O
O—CH3
~3500 cm-1
H
H
O
O
O
Si
Al
Si
Si
(b)
H3C
H
~3580 cm-1
O
O—CH3
~3500 cm-1
H
H
O
O
O
Si
Al
Si
Si
(c)
H3C
CH3
~3580 cm-1
O
O—H
~3500 cm-1
H
H
O
O
O
Si
Al
Si
Si
(d)
H3C
CH3
O
O—H
H
H
O
O
O
Si
Al
Si
Si

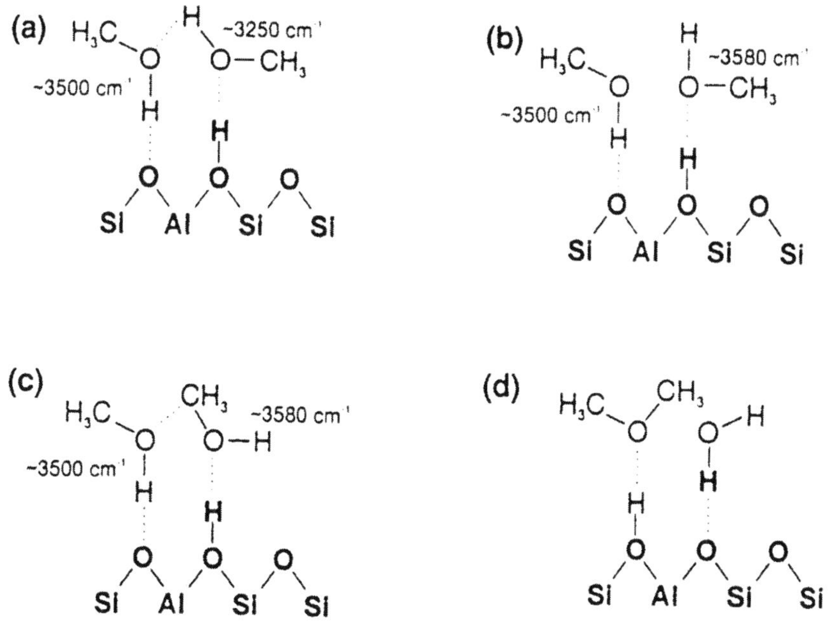

Fig. 6.8 Bimolecular mechanism proposed for the formation of dimethyl ether from two adjacent methanol molecules in HY. Reproduced with permission from [34].

tion of the two adsorbed methanol molecules so that dimethyl ether formation can occur, as shown in Fig. 6.8.

This pathway for dimethyl ether formation (a simultaneous bimolecular reaction rather than sequential formation of methoxy groups from one methanol followed by reaction of a second methanol with the methoxy groups) is supported by density functional calculations, which indicate that the bimolecular route has a lower activation barrier than the sequential route [51].

The mechanism proposed by Bonn et al. for dimethyl ether formation from methanol under catalytic conditions (thermal dissociation of the hydrogen bond between two adjacent adsorbed methanol molecules allowing reorientation to form dimethyl ether) is not completely consistent with the results of pulsed in situ FTIR studies of methanol in ZSM-5 [44, 46]. In particular, injection of pulses of methanol into HZSM-5 at 423 K does not give the infrared band at 3250 cm^{-1} attributed to the hydrogen bond between two co-adsorbed methanol molecules in HY. The three ν(OH) bands observed (Fig. 6.6) at around 3000, 2400, and 1700 cm^{-1} can all be accounted for in terms of strong hydrogen bonding of a single methanol molecule to Brønsted acid hydroxyl groups of the zeolite. Of course, the local concentrations of methanol in the zeolite pores in the pulsed microreactor experiment will be lower than those in the static in situ experiments of Bonn et al. (where catalyst samples were exposed to a static pressure of methanol above 10^{-2} mbar), and it may be argued that the steady-state concentrations of the doubly-adsorbed methanol species are too low to detect in the pulsed experiment. Nevertheless, it is clear from the pulsed experiments that the major adsorbed species produced from methanol and dimethyl ether, respectively, in HZSM-5 at low coverages and 423 K are quite distinct from each other; both form the corresponding molecular species strongly hydrogen-bonded to the Brønsted acid sites [46].

The infrared spectra obtained on injection of methanol or dimethyl ether into HZSM-5 at 523 K are quite different from those obtained at lower temperatures. For example, Fig. 6.9 shows difference spectra measured following injection of a single pulse of dimethyl ether into a high aluminum content HZSM-5 at 423 K, 473 K, and 523 K.

The spectrum at 423 K is that of hydrogen-bonded dimethyl ether. At 523 K, the spectrum in the ν(CH) region is dominated by a single pair of bands due to methoxy groups formed at Brønsted acid sites:

$$CH_3OCH_3 + Si(OH)Al \rightarrow Si(OCH_3)Al + CH_3OH$$

This identification [44] was confirmed by carrying out similar experiments in HZSM-5 zeolites containing different concentrations of Brønsted acid sites. A closely similar spectrum was obtained when methanol was injected into the same zeolite at the same temperature:

$$CH_3OH + Si(OH)Al \rightarrow Si(OCH_3)Al + H_2O$$

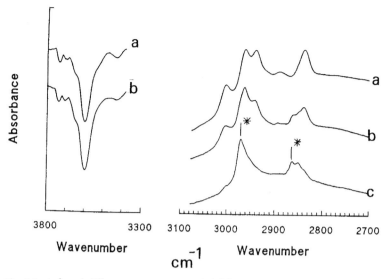

Fig. 6.9 Infrared difference spectra recorded following injection of dimethyl ether into HZSM-5 (Si:Al = 12) at (a) 423 K, (b) 473 K, and (c) 523 K. Asterisks denote bands assigned to methoxy groups formed at Brønsted acid sites. Reproduced with permission from [46].

The conversion of methanol or dimethyl ether, respectively, into surface methoxy groups is an activated process that only occurs at 473 K or above. This observation is consistent with the density functional calculations of van Santen et al. [51]. By measuring reaction products down stream from the infrared cell, it was shown that the first appearance of the Brønsted site bound methoxy groups in the infrared spectrum correlated with the first appearance of hydrocarbon products (light alkenes) [44, 46], and the onset of isotopic exchange between CD_3OH and zeolite hydroxyl groups [44].

A further observation suggesting that the methoxy groups play an important role in the methanol-to-gasoline chemistry is that they are active methylating agents. Injection, for example, of a pulse of benzene into the in situ cell following formation of the methoxy groups from methanol or dimethyl ether generates toluene, and decreases the intensity of the methoxy bands in the infrared spectrum [44].

The in situ infared experiments do not, however, answer the underlying fundamental question of how the first carbon-carbon bonds are formed from methanol or dimethyl ether in the MTG chemistry [43]. Using conventional infrared spectroscopy, it is not possible to directly observe short-lived reaction intermediates. Improvements in signal-to-noise in the time-resolved experiment offer some real hope of being able to do this, however. Selectively exciting the $\nu(CH)$ vibrations of the methoxy groups, for example, may answer the question as to whether a surface-bound ylide species, OCH_2^-, is a key intermediate, or that of whether trimethyloxonium species have a transient existence under the reaction conditions.

The importance of bimolecular reaction pathways involving two or more methanol or dimethyl ether molecules under the reaction conditions also needs clarification. Under conditions of working catalysis, an excess of methanol or dimethyl ether will be present, and bimolecular reaction pathways involving clusters of two or more hydrogen-bonded methanol or dimethyl ether molecules may be more significant than the sequential processes identified in the pulsed-flow microreactor studies described above.

6.3.1.3 Side-chain alkylation of toluene

The side-chain alkylation of toluene with methanol over basic catalysts to produce styrene is of considerable interest as a potential new route to this important chemical. It is known that basic zeolites such as Cs- and Rb-exchanged X zeolite catalyze this reaction, whereas more acidic zeolites produce ring-alkylated products, i.e. xylenes. Palomares et al. [52] have recently described an in situ infrared spectroscopic study of this reaction over Cs and Rb X and Cs Y zeolites, and compared it with the reaction over basic MgO and hydrotalcite.

Transmission spectra were collected using the stirred tank reactor cell described in [8]. Because the conversions achieved with the amount of catalyst present in a typical pressed disk for infrared spectroscopy were so low, parallel experiments were also performed with a tubular quartz reactor in place of the infrared cell in the experimental set up.

Methanol alone adsorbs on alkali metal exchanged zeolites in three different ways: through a Lewis acid-Lewis base type of interaction of the methanol oxygen with the cation, through hydrogen-bonding to zeolite framework oxygens, and through hydrogen-bonding with other methanol molecules [53]. Toluene alone was found to adsorb molecularly, with only slight frequency shifts between the different catalysts. Striking differences were found, however, between the catalysts in co-adsorption experiments. As shown in Fig. 6.10, the relative amounts of methanol and toluene found on the surface when the catalysts were exposed to equal pressures of both reactants varied from 97% methanol on MgO to 75% toluene on CsY zeolite.

In the parallel reactor studies, it was found that only CsX gave side-chain methylated products (around 80% selectivity). CsY gave high selectivity in ring alkylation (83%), while MgO and hydrotalcite gave only decomposition products (formaldehyde, dimethyl ether, CO, CO_2, H_2, and H_2O). Difference spectra of CsX obtained during reaction between 423 K and 773 K showed that as the temperature was raised the bands due to adsorbed toluene were removed, but a band assigned to adsorbed formate species at 1610 cm^{-1} was also detected. This band was not seen in the corresponding spectra of CsY.

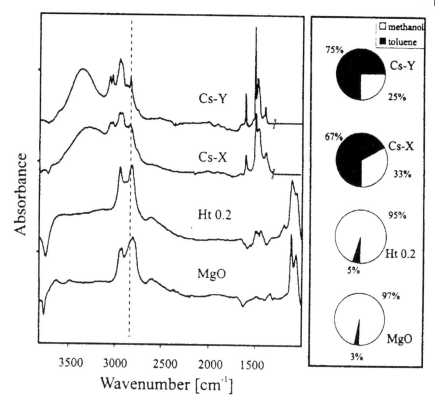

Fig. 6.10 Difference infrared spectra and relative concentrations of adsorbed methanol and toluene following exposure of different catalysts to equal pressures of methanol and toluene at 308 K. Reproduced with permission from [52].

From these in situ infrared studies coupled with catalytic measurements, the authors concluded that a zeolite catalyst for the side-chain alkylation of toluene must meet three requirements: to establish an adequate adsorption stoichiometry between toluene and methanol, to have sufficient base strength to dehydrogenate methanol to formaldehyde, and to stabilize and activate the methyl group of toluene within the pores. A reaction pathway involving aldol-type condensation was proposed.

This study nicely illustrates the power of in situ infrared spectroscopy to observe steady-state concentrations of adsorbed species under reaction conditions.

6.3.1.4 **Selective Catalytic Reduction (SCR) of NO$_x$**

The use of metal-exchanged zeolites to catalyze the selective reduction of NO$_x$ by hydrocarbons in the presence of oxygen has attracted widespread attention over

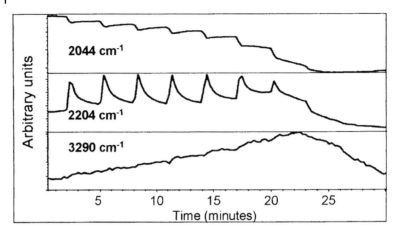

Fig. 6.11 Responses of the adsorbed species formed from NO + C_3H_8 over CuZSM-5 at 623 K to the introduction of successive pulses of oxygen. The 2044 cm^{-1} band is due to isocyanide, the 2204 cm^{-1} band to isocyanate, and the 3290 cm^{-1} band to ammonia. Reproduced with permission from [54].

the past ten years. Poignant et al. [54] have recently described an in situ infrared study of the selective reduction of NO_x by propane over CuZSM-5 zeolite catalysts, using an in situ transmission cell that allowed either continuous feed or pulse operation [55].

Infrared spectra recorded in the presence of flowing reactants (NO + propane + oxygen) gave a complex set of bands associated with partial oxidation products of propane (acrylates and carboxylates), plus additional multiple bands above 2000 cm^{-1} due to various other reaction intermediates. In the absence of oxygen, a strong band developed at 2044 cm^{-1}, attributable to a Cu^+ isocyanide species, CuNC. This species is stable under these conditions, but reacts when pulses of oxygen are subsequently introduced, forming bands characteristic of adsorbed ammonia (3365, 3290, 3193, and 1611 cm^{-1}). The progress of this reaction was followed by measuring spectra sequentially after the addition of successive pulses of oxygen. Figure 6.11 shows the time dependence of the 2044 cm^{-1} band, an ammonia band at 3290 cm^{-1}, and an intermediate band at 2204 cm^{-1} assigned to isocyanate, CuNCO.

These data show that the initial step in the process is the reaction of the isocyanide species with oxygen to form isocyanate. This, however, reacts further with traces of water to form ammonia, which is then oxidized to nitrogen.

$$2\ CuNC + O_2 \rightarrow 2\ CuNCO$$

$$2\ CuNCO + 3\ H_2O \rightarrow 2\ CuNH_3 + 2\ CO_2 + 1/2\ O_2$$

$$2\ CuNH_3 + 3/2\ O_2 \rightarrow 2\ Cu^+ + N_2 + 3\ H_2O$$

This scheme was supported by mass spectrometric analyses of the gas emitted from the infrared cell, which showed CO_2 and N_2 pulses being expelled from the cell following the introduction of oxygen pulses.

Under SCR conditions, NO is present as well, and under these circumstances ammonia reacts with NO according to:

$$2\ Cu(NH_3) + 2\ NO + 1/2\ O_2 \rightarrow 2\ Cu^+ + 2\ N_2 + 3\ H_2O$$

The SCR reaction is a complex one, and there are many issues remaining unresolved. With different hydrocarbon reactants and different catalysts, the reaction pathways may be different [55]. Nevertheless, in situ infrared experiments of the type described by Poignant et al., in which both continuous-flow and pulse experiments are coupled with on-line analysis of the reaction products, will play a major role in identifying the important reaction steps.

6.3.2
Oxide Catalysts

Although transmission infrared measurements have been historically the preferred method for obtaining spectra of oxide catalysts and adsorbed species, not all oxide catalysts lend themselves to the formation of thin infrared transparent disks, and diffuse reflectance methods have recently become more popular. In this section, examples of a transmission study, a diffuse reflectance study, and a picosecond time-resolved study are presented.

6.3.2.1 Selective catalytic reduction of NO by ammonia over vanadia/titania

The selective catalytic reduction of NO by ammonia over vanadia/titania catalysts in the presence of oxygen has been the subject of many studies, and a range of different reaction mechanisms has been proposed [56]. In an attempt to resolve this mechanistic uncertainty about a reaction of considerable industrial importance, Topsøe et al. [57, 58] combined in situ FTIR studies under both transient and steady-state conditions with on-line activity measurements.

Figure 6.12 shows in situ spectra recorded during temperature programmed surface reaction experiments (TPSR), in which a 6% vanadia/titania catalyst was first exposed to saturation coverage of ammonia, then heated from 325 K to 625 K in 50 K steps in flowing O_2, NO + O_2, or NO.

Adsorption of ammonia at room temperature on the oxidized catalyst produces bands due to two forms of chemisorbed ammonia: NH_3 coordinated to Lewis acid sites and NH_4^+ formed at Brønsted acid sites. The Brønsted sites were shown to be VOH groups on the vanadia component of the catalyst. On heating in flowing

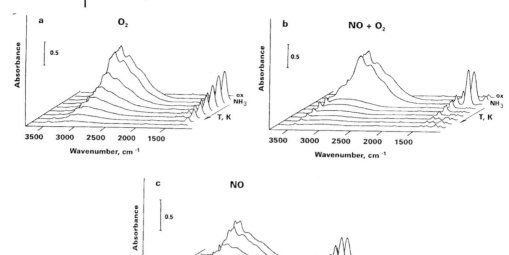

Fig. 6.12 In situ infrared spectra of 6% V_2O_5/TiO_2 catalyst following preadsorption of ammonia and subsequent stepwise heating from 375 K to 625 K at 50 K intervals in different gas streams. Reproduced with permission from [57].

oxygen (Fig. 6.12(a)), both forms of adsorbed ammonia were gradually removed, and the VOH groups (a band at 3640 cm^{-1}) were restored. Downstream analysis of the gas phase revealed that desorption of ammonia was the major process occurring, although some N_2 and H_2O due to ammonia oxidation were detected at higher temperatures. No oxides of nitrogen were detected during TPSR in oxygen.

When the preadsorbed ammonia was heated in the presence of an oxygen/nitric oxide mixture, substantially different spectra were obtained (Fig. 2.12(b)). The bands due to NH$_4^+$ were initially enhanced and were accompanied by bands due to adsorbed water. Downstream gas-phase analysis showed that high concentrations of N_2 and H_2O formed even at room temperature. At 375 K, the bands due to adsorbed ammonia were completely removed, and the VOH band was restored (initially shifting to higher frequency, then returning to its original frequency). Above 400 K, significant yields of NO_2 were observed in the gas phase, but no infrared bands that could be attributed to adsorbed oxides of nitrogen were detected.

Heating preadsorbed ammonia in nitric oxide alone (Fig. 6.12(c)) caused a similar initial enhancement in the bands due to NH$_4^+$, accompanied by adsorbed water bands, but in this case the adsorbed ammonia declined more slowly on heating to higher temperatures, and this was accompanied by restoration of the VOH groups. Gas-phase analysis detected only nitrogen and water.

Topsøe et al. also monitored changes in the infrared bands due to overtones of V=O vibrations of the catalyst during TPSR of ammonia (these bands are too weak to be clearly visible on the intensity scale used in Fig. 6.12). Ammonia adsorption shifts the main V=O overtone band to lower frequency by about 80 cm^{-1}. TPSR in NO + O$_2$ quickly restores this band to its original position and intensity, whereas with O$_2$ and NO individually, the frequency shift occurs more gradually over a much wider temperature range.

Under steady-state SCR reaction conditions (flowing ammonia, nitric oxide, and oxygen at 525 K), in situ FTIR spectra showed that adsorbed ammonia species (Lewis acid and Brønsted acid bound) were dominant on the surface, although at lower concentrations than when ammonia alone was adsorbed at room temperature. Under the reaction conditions, VOH groups were also observed, but the V=O overtone band of the catalyst at 2040 cm^{-1} was totally shifted to lower frequency.

To relate the various surface species and spectral changes observed under in situ continuous-flow SCR conditions, Topsøe et al. correlated the intensities of the various infrared bands with the NO$_x$ conversion as determined by downstream product analysis. Figure 6.13 shows plots of NO$_x$ conversion versus band intensities (measured in situ) for different catalysts (different vanadia loadings) and different oxygen contents of the reaction stream [58].

It is clear from these correlations that it is the Brønsted acid site bound ammonia that is reactive in the SCR reaction. The number of Brønsted acid sites is proportional to the vanadia loading of the catalyst; furthermore, the V=O species responsible for the 2025 cm^{-1} overtone band shows the best correlation with catalyst performance.

From these and other observations on this system, the authors were able to put together a catalytic cycle for the SCR reaction as it operates under industrial conditions. The reaction is initiated by ammonia adsorption on Brønsted acid sites, which are VOH groups. The adsorbed ammonia is activated by interaction with adjacent V=O groups, which involves partial transfer of a hydrogen atom, leading to reduction of V(V) to V(IV). Either gaseous or more likely weakly adsorbed NO then reacts with the adsorbed ammonia to yield N$_2$ and H$_2$O and releasing V(V)OH and V(IV)OH sites, which are then reoxidized to V=O by O$_2$, thereby completing the cycle.

This study of a complex reaction system illustrates the power of in situ infrared spectroscopy to observe both adsorbed species and surface sites (V=O and VOH) under actual reaction conditions, and, when coupled with on-line product analysis, to deduce a reaction pathway that is completely consistent with the global reaction kinetics.

Another example of an application of the in situ transmission technique is the study by Burcham and Wachs [59] of methanol oxidation over supported vanadia catalysts. The approach taken by these authors was very similar to that of Topsøe et al.: in situ measurements of adsorbed species, surface hydroxyl groups, and V=O

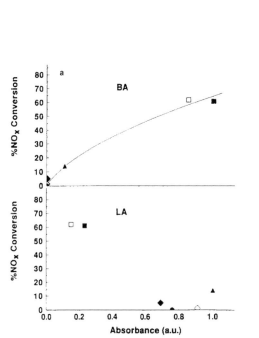

Fig. 6.13 Correlation of NO$_x$ conversion with infrared band intensities measured in situ during reaction of NO + NH$_3$ + O$_2$ over different vanadia/titania catalysts at 525 K (O$_2$ partial pressure also varied). BA is NH$_4^+$, LA is Lewis acid site bound ammonia, and the bands at 2040, 2025, and 1970 cm^{-1} are due to different V=O overtone vibrations. Reproduced with permission from [58].

overtone vibrations were coupled with product stream analysis to investigate the effect of catalyst support on reaction pathways.

6.3.2.2   In situ DRIFTS study of NO reduction by CH$_4$ over La$_2$O$_3$

Many metal oxide catalysts have low surface areas and particle morphologies that make the preparation of thin pressed disks with sufficient surface area to give reasonable sensitivity toward adsorbed species in transmission FTIR measurements difficult or impossible. Huang et al. have recently reported overcoming these problems with a lanthana catalyst by using the in situ diffuse reflectance technique (DRIFTS) [60].

Rare earth oxides are known to be effective catalysts for the SCR of NO with CH$_4$ at temperatures above 800 K [61]. Huang et al. employed La$_2$O$_3$ with a surface area of only 2.4 m² g⁻¹ and used a commercial DRIFTS in situ cell to observe spectra of the powdered catalyst in different reactant gas streams at 800 K. NO alone at this temperature gave bands due to adsorbed NO^- or $(N_2O_2)^{2-}$ at around 1120 cm⁻¹. These bands were totally absent when the same catalyst was exposed to $NO + CH_4$ mixtures at 800 K; in this case, adsorbed carbonates were the dominant surface species, but a weak band due to adsorbed CN was also observed. Figure 6.14 shows spectra measured in the presence of either $NO + O_2$ or the full SCR reaction mixture. In both cases, the spectrum of the freshly calcined catalyst has been subtracted so that only spectral changes on exposure to the reactant gases are recorded.

Exposure of the catalyst to $NO + O_2$ gave strong bands at 1540, 1262, and 1006 cm⁻¹, which are due to a unidentate nitrate species NO_3^- on the surface. Also present are bands due to gas-phase NO and NO_2 (Fig. 6.14(b)). In the presence of

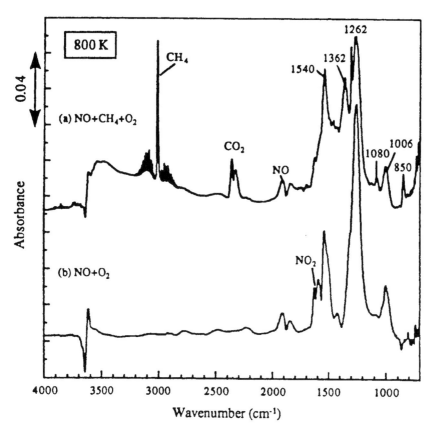

Fig. 6.14 In situ DRIFT spectra recorded during the flow of different gas mixtures over La_2O_3 at 800 K. Bands due to gas-phase species are designated. Reproduced with permission from [60].

CH_4 as well (Fig. 6.14(a)), the nitrate bands are still observed, but additional bands due to the same unidentate carbonate species as seen for $NO + CH_4$ mixtures appear (1540, 1362, 1080, and 850 cm^{-1}), as well as bands due to gas-phase CO_2.

From these in situ DRIFTS studies and parallel TPR and TPD experiments (not carried out in the DRIFTS cell), the authors concluded that in the absence of oxygen, NO reduction occurs by reaction of NO^- or $(N_2O_2)^{2-}$ species with adsorbed methane. The unidentate NO_3^- species formed in the presence of oxygen does not appear to react directly with methane. Since formation of the $NO^-/(N_2O_2)^{2-}$ species is suppressed in the presence of oxygen, it is suggested that the NO reduction pathway is different under these conditions; either an adsorbed O species or an NO_2^- species may abstract a hydrogen atom from CH_4 to form CH_3 radicals, although no direct spectroscopic evidence for this pathway has yet been obtained.

The quality of spectral data obtained in this initial in situ DRIFTS study indicates that the technique has considerable potential for studying reaction pathways over low surface area oxide catalysts, particularly if coupled with on-line product analysis.

6.3.2.3 Picosecond Infrared Spectroscopy on Single-Crystal Oxide Surfaces

The formation and reaction of formate species are important steps in catalytic reactions over many different surfaces. Domen et al. [31] have recently applied the picosecond time-resolved sum-frequency generation technique to observe the dynamics of formate species on an NiO(111) single-crystal surface. In particular, they were able to observe the thermally induced transformation of the stable bidentate formate species formed on this surface on exposure to formic acid to a monodentate species that initiates the decomposition reaction. Figure 6.15 shows a schematic representation of the time-resolved experiment performed.

The surface of the crystal containing adsorbed formate is rapidly heated to about 300 K above the initial temperature with a 35 ps near-infrared pump pulse. After a variable delay, one point in the spectrum of the adsorbed phase is obtained with an SFG pulse; the SFG intensity generated by a particular combination of variable infrared and fixed visible laser pulses is measured. The system is allowed to relax for 0.1 s; an ambient pressure of 10^{-5} Pa of formic acid vapor restores any formate lost from the surface. A second pump pulse is then applied, and the SFG response is probed with a different infrared frequency. This process is continued until a complete scan of the SFG spectrum is obtained for a given fixed delay between the pump and probe pulses. The delay time between the pump and probe pulses is then altered, and the experiment is repeated. In this way, it is possible to build up a set of spectra of the adsorbed species at various delay times after the near-infrared heating pulse has been applied.

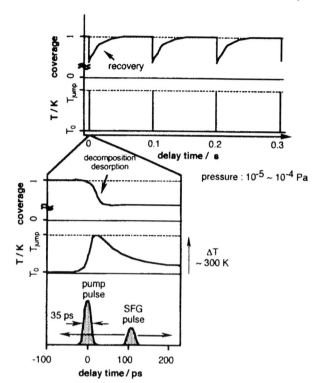

Fig. 6.15 Schematic of
the time-resolved
infrared experiment
used by Domen et al.
to study adsorbed
formate species on
NiO(111). Reproduced
with permission from
[31].

When formed from [D₂]formic acid, the stable bidentate adsorbed formate species has a characteristic C–D stretching vibration at 2160 cm⁻¹. When the surface was irradiated with the near-infrared pump pulse, this band weakened and a new band appeared at 2190 cm⁻¹, which was assigned to the C–D stretching vibration of a monodentate formate species. Figure 6.16 shows the time dependence of the two bands following application of the heating pulse.

It is clear that the loss of the bidentate species on application of the heating pulse correlates directly with the appearance of the monodentate species. Subsequent decay of the monodentate species as the surface cools restores most but not all of the bidentate species.

Domen et al. interpret these data in terms of a simple kinetic scheme, in which the bidentate and monodentate formate species are in equilibrium, and decomposition of the monodentate species leads to reaction products.

$$DCOO_{bident} \leftrightarrow DCOO_{monodent} \rightarrow D_2, CO_2, D_2O, CO$$

The kinetic parameters extracted for this mechanism from the time and temperature dependence of the SFG signals are in approximate agreement with those de-

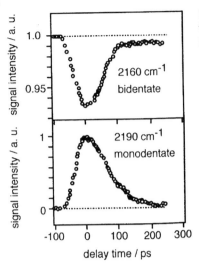

Fig. 6.16 Time profiles of SFG signals from formate species on NiO(111) at 400 K. Reproduced with permission from [31].

duced from conventional kinetic analysis of the formic acid decomposition reaction over NiO(111).

The power of this in situ time-resolved spectroscopic method for directly observing unstable intermediate species on a catalyst surface is very evident. In the same paper, Domen et al. also presented preliminary data for methoxy groups formed on an Ni(111) surface [31].

6.3.3
Supported Metal Catalysts

Both transmission and diffuse reflectance techniques have been used to measure in situ infrared spectra of supported metal catalysts. The choice of method depends on the support material used. Zeolite and high surface area oxide supports are conveniently studied as pressed disks; in other cases, such as carbon supports, diffuse reflectance may be the only viable sampling method. Three examples are described here.

6.3.3.1 **Alkane reactions over bifunctional zeolites**

Alkane isomerization over metal-loaded zeolite catalysts is an important process in the production of environmentally acceptable gasoline components. Höchtl et al. have described an in situ infrared study using transmission of *n*-heptane and 2-methylhexane conversion over platinum-loaded HZSM-5 and HBETA zeolites

[62]. A stirred tank reactor in situ cell [8] was used, and on-line product analysis by gas chromatography was employed. During n-heptane conversion over Pt-HZSM-5 at 483 K, n-C_7 was the only adsorbed species detected. The spectra obtained at this temperature showed that the adsorbed hydrocarbon interacted weakly with the Brønsted acid hydroxyl groups in the zeolite, and no bands due to adsorbed alkenes could be detected. At higher reaction temperatures, the surface coverage of adsorbed hydrocarbon was reduced, restoring the Brønsted hydroxyl groups, and the spectrum in the C–H stretching region changed from that characteristic of n-heptane to one that could be fitted to a sum of the spectra of propane and isobutane.

Figure 6.17 shows time-resolved spectra of Pt-HZSM-5 at 523 K measured after the flow of n-heptane to the reactor had been stopped. The restoration of the Brønsted acid sites and the conversion of adsorbed n-heptane to adsorbed propane and isobutane can be seen to occur over a time scale of up to 10 minutes at this temperature.

In these experiments, the adsorbed species detected by in situ FTIR correlated very well with the reaction products (principally propane and isobutane) detected downstream by gas chromatography. Under the conditions employed, intermediates of the isomerization process could not be detected.

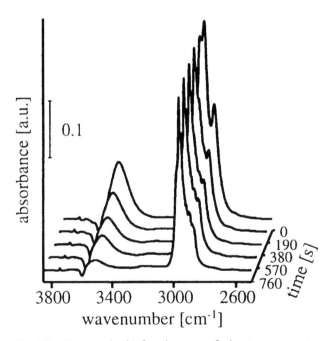

Fig. 6.17 Time-resolved infrared spectra of n-heptane conversion over PtZSM-5 at 523 K following interruption of the n-heptane flow. Reproduced with permission from [62].

6.3.3.2 **DRIFTS study of NO decomposition over carbon-supported Rh and Pd**

Carbon supports are commonly employed in practical catalysis because of their high surface areas and chemical inertness. However, such materials are totally intractable for transmission infrared studies, and diffuse reflectance is the only sampling technique available for in situ studies of carbon-supported catalysts. Recently, Almusaiteer et al. have described in situ DRIFTS studies of carbon-supported Rh and Pd catalysts for NO decomposition [63]. The catalytic decomposition of NO is thermodynamically favored, but most catalysts show low activity because of the poisoning of active sites by strongly adsorbed oxygen derived from the dissociated NO [64]. Carbon as a support offers the attractive possibility of facilitating oxygen removal as CO_2. Almusaiteer et al. used a commercial in situ DRIFTS cell containing the carbon-supported catalyst diluted 10× with KBr to achieve sufficient signal-to-noise in the diffuse reflectance spectra, which was connected in series with an additional reactor containing further catalyst. Effluent from both reactors was analyzed by mass spectrometry.

Figure 6.18 shows diffuse reflectance spectra measured in such an experiment with an Rh/C catalyst following the introduction of NO at 673 K. The major species present are carbon-oxygen groups on the surface of the carbon support (bands at 1402 and 1277 cm^{-1}), although traces of carbon-nitrogen species are also detected.

The gas-phase products detected downstream from the infrared cell during these experiments were principally N_2 and CO_2. In particular, the onset of CO_2 production coincided with the steep increase in the 1402 cm^{-1} infrared band due to surface carbon-oxygen species on the carbon support. Pd/C catalysts, on the other hand, showed no evidence for the formation of surface carbon-oxygen species on the support under similar circumstances. The spectra in this case showed bands attributable to CN species on the support and Pd–NO species, while the gas-phase analysis showed steady-state production of oxygen.

The signal-to-noise in the DRIFTS spectra of the carbon-supported metal catalysts in [63] is low, and the bands due to adsorbed species are poorly defined. Nevertheless, the in situ observation of the surfaces of these catalysts does allow a rationalization of the differences in catalytic performances of Rh/C and Pd/C catalysts for NO decomposition, and the differences between these catalysts and the corresponding alumina-supported catalysts can be rationalized in terms of the different capabilities of the catalysts to manipulate adsorbed oxygen [63].

6.3.3.3 **DRIFTS study of NO_x reduction by propene**

The possible consequences for diffusion of reactants and products of pressing oxide supports into disks for transmission spectroscopy have prompted some work-

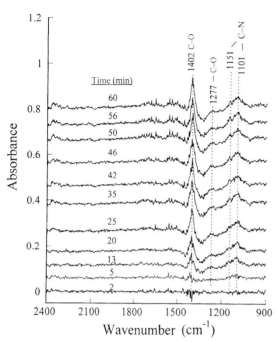

Fig. 6.18 DRIFTS spectra of Rh/C catalyst following introduction of an NO flow at 673 K. Reproduced with permission from [63].

ers to use diffuse reflectance even with materials that are readily studied by transmission. For example, Okuhara et al. have used an in situ DRIFTS technique to study propene reduction of NO_x over a Pt/SiO_2 catalyst [65]. A commercial DRIFTS cell was connected in series to a tubular reactor and the effluent was analyzed with a high-speed gas chromatograph.

In situ observation of spectra under continuous-flow conditions (NO_2 or $NO + C_3H_6 + O_2$) revealed the major adsorbed species at low temperatures (below 373 K) to be organonitro and nitrite species (the identification of these bands was confirmed by observing isotope shifts with ^{15}NO). At higher temperatures, carbonyl compounds were detected on the surface (a band at 1740 cm^{-1} was unshifted when ^{15}NO was employed), but above 473 K the major surface species became isocyanate (a band at 2174 cm^{-1}, which shifted by the expected amounts when ^{15}NO and $N^{18}O + ^{18}O_2$ were used). The significance of each of these species to the SCR reaction was investigated by observing the response of each band to changes in the input gas stream composition while simultaneously monitoring the product gas composition.

Figure 6.19 shows the responses of the organonitro species and the isocyanate groups to changes in input gas stream composition at 393 K and 433 K, respectively. In each case, a steady-state concentration of the adsorbed species was established by passing $NO_2 + C_3H_6 + O_2$ or $NO + C_3H_6 + O_2$ over the catalyst at the appro-

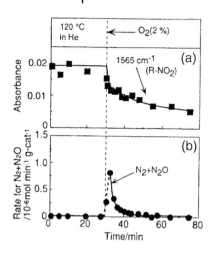
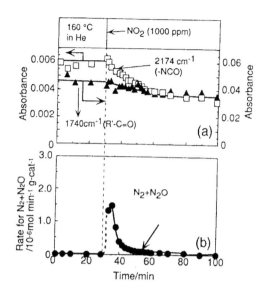

Fig. 6.19 Responses of infrared bands from a Pt/SiO₂ catalyst in an NO$_x$ + C$_3$H$_6$ + O$_2$ stream to changing gas composition (at 393 K and 433 K, respectively). Reproduced with permission from [65].

priate temperature; the gas flow was then switched to helium, and then changed to either O$_2$ or NO$_2$.

The data show that at 393 K the organonitro species react with O$_2$ to form N$_2$ and N$_2$O. Similar results were obtained with NO$_2$, but the organonitro species did not react with NO or with propene. At 433 K, the carbonyl species are unreactive towards NO$_2$, but the isocyanate species does react with NO$_2$ at this temperature to form N$_2$ and N$_2$O.

Figure 6.20 shows the response of the isocyanate species at 453 K to changing the composition of the feed gas from NO + C$_3$H$_6$ + O$_2$ to either NO alone or to NO$_2$ alone.

It is clear from these data that the isocyanate species is unreactive towards NO, but does react with NO$_2$ at this temperature. Similar reactivity towards oxygen was observed (not shown). On the other hand, the isocyanate species did not react with propene.

Okuhara et al. concluded from these measurements that two different mechanisms are operative for the SCR of NO or NO$_2$ with propene in the presence of oxygen over Pt/SiO$_2$. At low temperatures, propene reacts with NO$_2$ to form organonitro species, which are reaction intermediates in the formation of N$_2$ and N$_2$O under these conditions. The oxidation of NO to NO$_2$ is accelerated by platinum. At higher temperatures, isocyanate species are a key intermediate.

Many aspects of these reactions remain uncertain. In particular, it is not clear as to how the isocyanate species are formed (whether or not they derive from the orga-

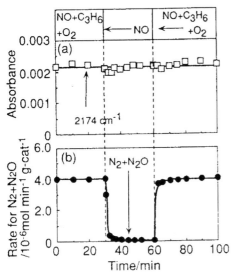

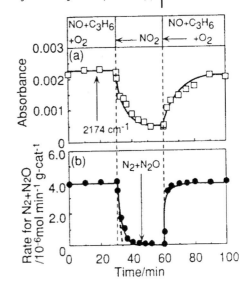

Fig. 6.20 Responses of infrared bands from a Pt/SiO$_2$ catalyst in an NO$_x$ + C$_3$H$_6$ + O$_2$ stream to changing gas composition at 453 K. Reproduced with permission from [65].

nonitro species). Nevertheless, the in situ DRIFTS experiment is able to identify the major adsorbed species present under the reaction conditions and to elucidate their relative reactivities; these are both key steps in deducing reaction mechanisms.

6.3.4
Metal Surfaces

Investigations of unsupported metal surfaces are usually confined to model studies aimed at elucidating possible reaction mechanisms. Both reflection-absorption infrared spectroscopy (RAIRS) and the more recent SFG technique have been applied to such studies, particularly with a view to overcoming the "pressure gap" between the usual surface science experiments and working catalysts. A second area of catalysis where in situ infrared spectroscopy can be applied is electrocatalysis; in this case, unsupported metal electrode surfaces are studied. Examples of both of these types of application are presented here to illustrate the procedures involved.

6.3.4.1 In situ RAIRS study of kinetic oscillations in the Pt(100) NO + CO system

Magtoto and Richardson [66] have recently described an in situ RAIRS investigation of the species present on a Pt(100) single-crystal surface during the reaction between NO and CO, which is well known for showing nonlinear behavior and kinetic oscillations [67]. The experiment was carried out in a conventional UHV chamber equipped for RAIRS measurements [68] coupled with a mass spectrometer for monitoring changes in gas-phase composition. At a typical reaction temperature of 470 K in a continuous flow of CO + NO, the rate of production of CO_2 oscillates uniformly with a temporal frequency of between 1 and 3 minutes, depending on the partial pressures of the reactants. Infrared spectra of the metal surface were collected at all points during the oscillations, typically every 7.8 s. Figure 6.21 shows a set of spectra collected over the 2.5 minute period of one oscillation.

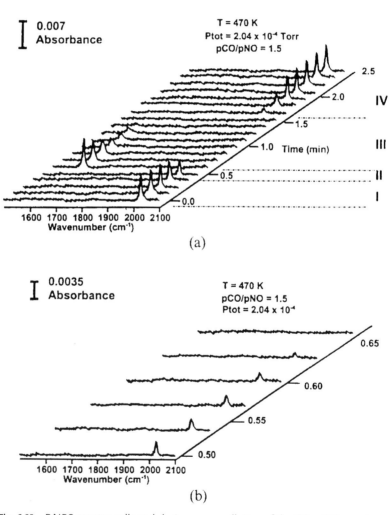

(a)

(b)

Fig. 6.21 RAIRS spectra collected during one oscillation of the CO + NO reaction over Pt(111) at 470 K. An expansion of the surface explosion region (II) is shown in (b). Reproduced with permission from [66].

At time $t = 0$, the rate of CO_2 formation is at a minimum, and the surface shows a high concentration of the linearly bonded adsorbed CO species with a vibrational frequency of 2030 cm^{-1}. Over the subsequent 30 seconds, this species decays, as the rate of CO_2 formation increases rapidly. After 50 seconds, the rate of CO_2 formation reaches a maximum and remains steady for a further 60 seconds. At this point, the 2030 cm^{-1} band has totally disappeared, and a new intense band is present at 1618 cm^{-1} due to chemisorbed NO, which gradually decays. The expansion of region II, the so-called explosion in CO_2 formation, in Fig. 6.21(b) shows that the 1618 cm^{-1} band does not appear until the adsorbed CO band is completely lost. In the final part of the cycle, as the rate of CO_2 formation decays to its original value, the adsorbed NO band has completely disappeared, and the adsorbed CO band gradually reappears.

According to Fink et al. [69], the kinetic oscillations in the CO + NO reaction are controlled by the vacant site requirements for the dissociative adsorption of NO. Magtoto and Richardson account for the observed changes in the infrared spectra in terms of this model. When the rate of CO_2 formation rate is at a minimum, the surface is almost completely covered with adsorbed CO, which poisons the reaction. NO adsorption and dissociation occurs at the small number of available vacant sites (the concentration of adsorbed NO is too low to detect by infrared spectroscopy). As reaction occurs between adsorbed CO and adsorbed O formed from NO, the concentration of adsorbed CO starts to fall and the rate of formation of CO_2 increases. This leads to the surface explosion. Dissociation of NO occurs freely at the vacant sites generated, building up a layer of adsorbed oxygen on the surface. It is the chemisorption of NO onto this adsorbed O layer that is responsible for the 1618 cm^{-1} infrared band (adsorbed N atoms readily desorb as N_2). Gas-phase CO eventually begins to consume the adsorbed O layer to form CO_2, whereupon the concentration of adsorbed NO decays. The freed adsorption sites are preferentially populated by CO, restoring the 2030 cm^{-1} band and reducing the rate of formation of CO_2 to its minimum value; the whole cycle then repeats.

This in situ RAIRS study has allowed the observation of surface species under steady-state continuous-flow conditions with a time resolution of better than 10 seconds. Under the conditions of the experiment, synchronization of the kinetic oscillations is transmitted through gas-phase oscillations in the partial pressures of CO and NO.

6.3.4.2 RAIRS studies of electrocatalysis

In situ infrared studies of electrocatalysis provide a different set of experimental challenges from those in heterogeneous gas-phase catalysis. External reflectance methods are used, and cell designs are available which minimize the path of the infrared beam through the strongly absorbing solvent. The applications of infrared spectroscopy to study working single-crystal electrodes have been reviewed by Korzeniewski and Severson [70]. An example taken from their review is the electrochemical oxidation of ethylene glycol at a stepped Pt(335) single-crystal electrode surface. At negative potentials, dissociative chemisorption of ethylene glycol produces adsorbed carbon monoxide. The infrared spectrum of the adsorbed CO is sensitive to the sites at which adsorption occurs, as is well known in gas-solid surface chemistry. In electrocatalysis, there is particular interest in identifying not only the adsorbed species but also the immediate products of oxidation found in the aqueous phase adjacent to the electrode surface. Figure 6.22 shows the quantities of reaction products produced from oxidation of ethylene glycol over Pt(335) as a function of the electrode potential. These values were determined from the intensities of the corresponding infrared bands, taking into account the integrated molar absorptivity coefficients corrected for electric field enhancement effects and the pathlength of the infrared beam through the thin layer adjacent to the electrode.

As the potential is increased to about 0.25 V, the adsorbed CO poison is removed from the electrode surface as CO_2, opening up reaction pathways that lead to formation of oxalic acid, the major product of ethylene glycol electro-oxidation.

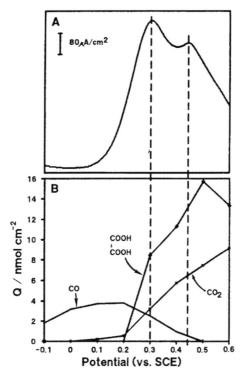

Fig. 6.22 Reaction products detected by in situ RAIRS observation of a Pt(335) electrode during electro-oxidation of ethylene glycol. The top curve is the corresponding electrovoltammogram. Reproduced with permission from [70].

6.4
Conclusions and Future Prospects

The conventional transmission infrared experiment is now an established technique with many applications to in situ studies. The importance of identifying the adsorbed species present on a working catalyst surface cannot be over-emphasized; any proposed reaction mechanisms must be able to account for the adsorbed species present. Identifying true reaction intermediates by conventional in situ infrared studies is, however, a harder problem. The majority of species present may well be spectators rather than direct participants. Observing the response of adsorbed species to changes in reactant gas composition and operating in situ infrared cells in pulse mode rather than continuous flow are all strategies which help to identify species that participate in the reaction pathway.

A second issue is the correct identification of adsorbed species from their vibrational frequencies. Where such species are strongly perturbed by the catalyst surface, relying on comparisons with model compounds may be dangerous, but the role of theory in predicting vibrational frequencies is rapidly growing in stature.

The experimental challenges for the conventional infrared experiment are to further improve sensitivity, shorten the time required to collect data, and to apply in situ methods to higher temperature reactions. The diffuse reflectance technique, although still suffering signal-to-noise problems in comparison with the transmission experiment, is growing in popularity as a method for collecting spectra under conditions closer to those prevailing in a typical catalytic reactor. In situ emission spectroscopy has a real potential for studies of high-temperature catalytic reactions that has scarcely been realised. Microspectroscopic techniques may have a future in the parallel screening of catalysts in high-throughput (combinatorial) experiments.

The most exciting developments at present in terms of in situ infrared studies of catalytic reactions come from the area of picosecond spectroscopy. This remains a difficult and expensive experiment. Nevertheless, the ability to probe individual adsorbed species on the time scale of a few vibrations and to identify which of them are precursors to true reaction intermediates and therefore essential participants in the reaction pathways offers real hope of realizing the dream of all catalytic chemists of understanding reaction mechanisms.

References

1 A. Terenin, *Zhur. Fiz. Khim. 14*, **1362** (1940).

2 R. P. Eischens, W. A. Pliskin, *Advan. Catal. 10*, **2** (1958).

3 N. Sheppard, D. J. C. Yates, *Proc. Roy. Soc. A238*, **69** (1956).

4 J. B. Peri, R. B. Hannan, *J. Phys. Chem. 64*, **1526** (1960).

5 G. Busca, *Catal. Today 27*, **323** (1996).

6 J. W. Niemantsverdriet, *"Spectroscopy in Catalysis"*, VCH, Berlin, 1995.

7 J. A. Lercher, V. Veefkind, K. Fajerweg, *Vibrational Spec. 19*, **107** (1999).

8 G. Mirth, F. Eder, J. A. Lercher, *Appl. Spec. 48*, **194** (1994).

9 R. Mariscal, H. R. Reinhoudt, A. D. van Langeveld, J. A. Moulijn, *Vibrational Spec. 16*, **119** (1998).

10 S. S. C. Chuang, M. A. Brundage, M. W. Balakos, G. Srinivas, *Appl. Spec. 49*, **1151** (1995).

11 R. F. Hicks, C. S. Kellner, B. J. Savitzky, W. C. Hecker, A. T. Bell, *J. Catal. 71*, **216** (1981).

12 S. H. Moon, H. Windawi, J. R. Katzer, *Ind. Eng. Chem. Fund. 20*, **396** (1981).

13 M. Komiyama, Y. Obi, *Rev. Sci. Instr. 67*, **1590** (1996).

14 A. Drochner, M. Fehlings, K. Krauss, H. Vogel, *Chem. Eng. Technol. 23*, **4** (2000).

15 D. H. Sullivan, W. C. Connor, M. P. Harold, *Appl. Spec. 46*, **811** (1992).

16 Th. Weber, J. C. Muijsers, J. H. M. C. van Wolput, C. P. J. Verhagen, J. W. Niemantsverdriet, *J. Phys. Chem. 100*, **14144** (1996).

17 A. M. Vasallo, P. A. Cole-Clarke, L. K. Pang, A. Palmisano, *Appl. Spec. 46*, **73** (1992).

18 G. L. Carr, G. P. Williams, *SPIE Conf. Proc. 3153*, **51** (1997).

19 F. Schüth, D. Demuth, B. Zibrowius, J. Kornatowski, G. Finger, *J. Am. Chem. Soc. 116*, **1090** (1994).

20 J. A. Lercher, C. Gründling, G. Eder-Mirth, *Catal. Today 27*, **353** (1996).

21 F. Marlow, D. Demuth, G. Stucky, F. Schüth, *J. Phys. Chem. 99*, **1306** (1995).

22 M. Nowotny, J. A. Lercher, H. Kessler, *Zeolites 11*, **454** (1991).

23 S. Popescu, S. Thomson, R. F. Howe, *Phys. Chem. Chem. Phys.*, in press (2000).

24 G. Müller, T. Narbeshuber, G. Mirth, J. A. Lercher, *J. Phys. Chem. 98*, **7436** (1994).

25 F. Qin, E. E.Wolf, *Catal. Lett. 39*, **19** (1996).

26 R. G. Greenler, *J. Chem. Phys. 44*, **310** (1966).

27 D. W. Goodman, *Ann. Rev. Phys. Chem. 48*, **43** (1997).

28 Y. R. Shen, *Surf. Sci. 299–300*, **551** (1994).

29 P. S. Cremer, X. Su, G. A. Somorjai, Y. R. Shen, *J. Molec. Catal. A131*, **225** (1998).

30 G. Rupprechter, T. Dellwig, H. Unterholt, H.-J. Freund, *Stud. Surf. Sci. Catal. 130*, **3131** (2000).

31 K. Domen, K. Kusafuka, A. Bandara, M. Hara, J. N. Kondo, J. Kubota, K. Onda, A. Wada, C. Hirose, *Stud. Surf. Sci. Catal. 130*, **365** (2000).

32 J. C. Ostruwsky, D. Raftery, R. M. Hochstrasser, *Ann. Rev. Phys. Chem. 45*, **519** (1994).

33 E. J. Heilweil, M. P. Casassa, R. R. Cavanagh, J. C. Stephenson, *Ann. Rev. Phys. Chem. 40*, **143** (1989).

34 M. Bonn, H. J. Bakker, K. Domen, C. Hirose, A. W. Kleyn, R. A. van Santen, *Catal. Rev. Sci. Eng. 40*, **127** (1998).

35 J. N. Kondo, S. Liqun, K. Domen, F. Wakabayashi, *J. Phys. Chem. B 101*, **5477** (1997).

36 J. N. Kondo, S. Liqun, K. Domen F. Wakabayashi, *J. Phys. Chem. B 101*, **9314** (1997).

37 J. N. Kondo, L. Shao, K. Domen, F. Wakabayashi, *Catal. Lett. 47*, **129** (1997).

38 J. N. Kondo, K. Domen, F. Wakabayashi, *J. Phys. Chem. B 102*, **2259** (1998).

39 J. N. Kondo, K. Domen, F. Wakabayashi, *Catal. Lett. 53*, **215** (1998).

40 E. Yoda, J. N. Kondo, F. Wakabayashi, K. Domen, Appl. *Catal. A 194–195*, **275** (2000).

41 J. N. Kondo, E. Yoda, M. Hara, F. Wakabayashi, K. Domen, *Stud. Surf. Sci. Catal. 130*, **2933** (2000).

42 C. D. Chang, *"Hydrocarbons from Methanol"*, Marcel Dekker, NY, 1983.

43 G. J. Hutchings, R. Hunter, Catal. *Today 6*, **279** (1990).

44 T. R. Forester, R. F. Howe, *J. Am. Chem. Soc. 109*, **3076** (1987).

45 G. Mirth, J. A. Lercher, M. W. Anderson, J. Klinowski, *J. Chem. Soc., Faraday Trans. 86*, **3039** (1990).

46 S. M. Campbell, X. Z. Jiang, R. F. Howe, *Microporous Mesoporous Mat. 29*, **91** (1999).

47 S. R. Blaszkowski, R. A. van Santen, *J. Phys. Chem. 99*, **11728** (1995).

48 E. L. Meijer, R. A. van Santen, A. P. J. Jansen, *J. Phys. Chem. 100*, **9282** (1996).

49 A. G. Pelmenschikov, R. A. van Santen, J. Jänchen, E. Meijer, *J. Phys. Chem. 97*, **11071** (1993).

50 M. Bonn, R. A. van Santen, J. A. Lercher, A. W. Kleyn, H. J. Bakker, *Chem. Phys. Lett. 278*, **213** (1997).

51 S. R. Blaszkowski, R. A. van Santen, *J. Am. Chem. Soc. 118*, **5152** (1996).

52 A. E. Palomares, G. Eder-Mirth, M. Rep, J. A. Lercher, *J. Catal. 180*, **56** (1998).

53 M. Rep, A. E. Palomares, G. Eder-Mirth, J. G. van Ommen, N. Rösch, J. A. Lercher, *J. Phys. Chem. B 104*, **8624** (2000).

54 F. Poignant, J. L. Freysz, M. Datari, J. Saussey, J. C. Lavalley, *Stud. Surf. Sci. Catal. 130*, **1487** (2000).

55 N. W. Cant, I. O. Y. Liu, *Catal. Today 63*, **133** (2000).

56 H. Bosch, F. Janssen, *Catal. Today 2*, **369** (1988).

57 N. Y. Topsøe, H. Topsøe, J. Dumesic, *J. Catal. 151*, **226** (1995).

58 N. Y. Topsøe, H. Topsøe, J. Dumesic, *J. Catal. 151*, **241** (1995).

59 L. J. Burcham, I. E. Wachs, *Catal. Today 49*, **467** (1999).

60 J. Huang, A. B. Walters, M. A. Vannice, *Catal. Lett. 64*, **77** (2000).

61 X. Zhang, A. B. Walters, M. A. Vannice, *Catal. Today 27*, **41** (1996).

62 M. Höchtl, Ch. Kleber, A. Jentys, H. Vinek, *Stud. Surf. Sci. Catal. 130*, **377** (2000).

63 K. Almusaiteer, R. Krishnamurthy, S. S. C. Chuang, *Catal. Today 55*, **291** (2000).

64 Y. Li, W. K. Hall, *J. Catal. 129*, **202** (1991).

65 T. Okuhara, Y. Hasada, M. Misono, *Catal. Today 35*, **83** (1997).

66 N. P. Magtoto, H. H. Richardson, *Surf. Sci. 417*, **189** (1998).

67 R. Imbihl, G. Ertl, *Chem. Rev. 95*, **697** (1995).

68 S. Hong, H. H. Richardson, *J. Vac. Sci. Tech. A11*, **1951** (1993).

69 Th. Fink, J. Dath, R. Imbihl, G. Ertl, *J. Chem. Phys. 95*, **2109** (1991).

70 C. Korzeniewski, M. W. Severson, *Spectrochim. Acta 51A*, **499** (1995).

7

In situ XAS Characterization of Heterogeneous Catalysts

George Meitzner

7.1
Introduction: X-ray Absorption Spectroscopy (XAS)

Heterogeneous catalysts comprise a broad array of materials, including metals, crystalline or amorphous oxides, sulfides, carbides, etc., and more or less intimate mixtures among phases. These are employed under a wide range of conditions of temperature, pressure, and atmosphere, and in liquid-phase, electrochemical, and photochemical reactors. This chapter is concerned with X-ray absorption spectroscopy (XAS), which is applicable to this broad slate of substances, under conditions representative of their commercial use, and provides information on the chemical state and physical environment of the absorber element.

X-ray absorption spectroscopy requires exotic instrumentation that is beyond the reach of many (or any) institutions. Equipment and experimental support are arranged through the synchrotron light sources, located in the US at Stanford University, Brookhaven National Laboratory, Argonne National Laboratory, Lawrence Berkeley Laboratory, Cornell University, and Louisiana State University. Access is through peer-reviewed proposals and is competitive, but in the author's experience experiments that are well conceived and well presented are eventually supported, without charge for work intended for publication. Descriptions of the facilities can be found at the respective web sites [1].

The historical background of XAS was reviewed by von Bordwehr [2]. Earliest applications to samples of catalytic interest were by Lewis [3] at the Texaco Research Center and by Van Nordstrand [4] then at the University of Tulsa. The first in situ applications were developed by Lytle et al. [5]. The evolving theory of XAS has been reviewed [6–8].

Experimental aspects [9] and in situ cell designs [10] have also been reviewed. X-ray absorption spectroscopy is applicable, in principle, to all the elements. Those lighter than Al have core binding energies in the UV region and must be studied in vacuo. Interferences are rare since binding energies are spread over a wide range (e.g. experimentally accessible K-edges range from 112 eV for Be to just above 29,000 eV, the approximate upper limit of available monochromatic X-rays, for Sn [11]). However, acquisition of adequate signal-to-noise can be a problem when a light element is present at a low concentration in a dense matrix, e.g. 1 wt.% V_2O_5/ZrO_2. In the more favorable case of a heavier element in a light matrix (e.g. Mo enzyme cofactor in aqueous solution), useful spectra have been measured from absorber concentrations well below 100 ppm.

Most heterogeneous catalysts are designed to maximize surface area and are therefore poorly crystalline. X-ray absorption spectroscopy is a source of physical structural information applicable to samples that do not diffract X-rays. The spectrum is measured in an energy range in which only one element in the sample absorbs, and the information in the spectrum describes the environment of atoms of only that element (Fig. 7.1). The matrix comprising the other elements in the catalyst, and the atmosphere and cell windows, are relatively transparent. This makes X-ray absorption spectroscopy well suited for in situ catalyst characterization.

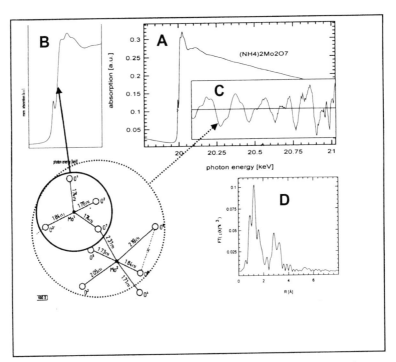

Fig. 7.1 The X-ray absorption spectrum (A) of $(NH_4)_2Mo_2O_7$ comprises the XANES spectrum (B), which describes the oxidation state and site symmetry, and the EXAFS (C), which describes the physical structure to a maximum range of about 1 nm. The Fourier transform of the EXAFS (D) is a radial distribution function describing the Mo sites in $(NH_4)_2Mo_2O_7$.

7.2
Information Content of XAS

7.2.1
X-ray Absorption Near-Edge Spectrum (XANES)

To measure a spectrum (field A of Fig. 7.1), the sample is illuminated by an X-ray beam while the X-ray energy is scanned and transmission, fluorescence intensity, or yield of secondary electrons is monitored. As the energy approaches a core-electron binding energy, a step-increase in absorbance occurs, referred to as the edge. The X-ray absorption near-edge spectrum (XANES) comprises that part of the spectrum within about 50 eV of the edge (Field B of Fig. 7.1).

7.2.1.1   Elemental analysis

The step height is proportional to the amount of absorber probed by the X-ray beam [12] according to the Beer–Lambert Law [13], and is most easily interpreted as an elemental analysis when the spectrum is measured in transmission mode. Even when the spectrum is measured by fluorescence, so that the penetration depth is not precisely known, the step height is useful to determine if absorber is being lost from the sample under in situ conditions. The concentration dependence of the step height, when the sampled volume is unknown, is also useful when an unknown concentration of one absorber can be determined by comparison of its step height with that of another absorber of known concentration in the same sample.

7.2.1.2   Oxidation state and site symmetry

The edge energy, equivalent to the binding energy, shifts in response to changes in oxidation state [14] (Fig. 7.2). By convention, edge energies are defined by the position of the first inflection point in the edge. Edge positions for the pure elements in their standard states have been tabulated by Bearden and Burr [11]. The K-edge shifts have typical magnitudes of about 2 eV per unit change in formal oxidation number. Spectral resolution is an order of magnitude better than this. However, electronic resonances that may occur at absorption edges can obscure the position of the transition to threshold. For this reason, shifts at transition metal L_{II} and L_{III} edges are small, since the p→d dipole transition, reflected in the resonance often called the white line, precedes the threshold transition in transition metals and their compounds. Fortunately, the oxidation state of the transition metal can usually be inferred from the intensity of the L_{II} or L_{III} edge white lines, which scale with the number of vacancies in the absorber d orbitals [16].

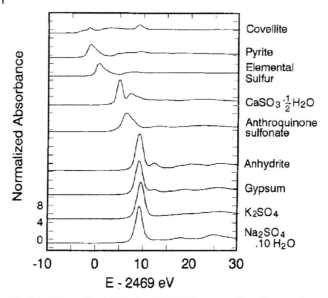

Fig. 7.2 Normalized S K-edge XANES from a series of inorganic compounds, referenced to the K-edge position of elemental sulfur [15]. The S 1s binding energy increases by about 12 eV on going from S^{2-} to S^{6+}.

In a narrow range below the binding (edge) energy, transitions may be excited to vacant states in the valence level that have appropriate symmetry (i.e. allowed transitions). When the 1s core level is excited in a K-edge spectrum, the dipole-allowed transition probes final states of p-type symmetry. Hybridization to molecular orbitals that lack an inversion center mixes p and d orbitals so that K-edge spectra from atoms in tetrahedral or distorted octahedral sites may include 1s to valence hybrid p-d molecular orbital vacancy transitions as pre-edge resonances [17].

7.2.1.3 Empirical analysis of XANES

In addition to elemental analysis, oxidation state, and site symmetry information, XANES provides a fingerprint of the average absorber atom in the sample. It is important to recognize that because the probing radiation is generally penetrating, XANES describes all the absorber in the spectroscopic sample regardless of crystallinity. Linear combinations of XANES from contributing species fitted to a sample spectrum [18] describe the fractions of absorber atoms in the corresponding sites (Fig. 7.3). Principle component analysis [20,21] can identify a species that occurs within a series of spectra, although always in admixture and the individual species is never recognizable by inspection of the spectra.

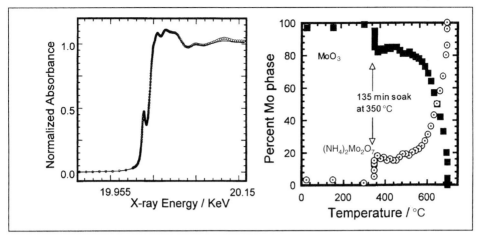

Fig. 7.3 Left: Representative fit to XANES spectrum from calcined Mo/H-ZSM5. The fit (circles) comprises 70% MoO₃ and 30% MgMo₂O₇-like Mo species. Right: Evolution of 4 wt.% Mo/H-ZSM5 under 20% O₂/He. MoO₃ progressively disperses within the zeolite channels as a ditetrahedral dimer resembling the MgMo₂O₇ oxyanion [19].

7.2.2
Extended X-ray Absorption Fine Structure (EXAFS)

Beyond the stepwise increase in absorbance marking the threshold for core electron excitation, oscillations in the value of the absorption coefficient are called extended X-ray absorption fine-structure (EXAFS; field C in Fig. 7.1). Photoelectrons are generated by absorption of X-rays, with the de Broglie wavelength given by:

$$\lambda_e = h/(2\,m_e\,E)^{1/2}$$

where

λ_e	= the de Broglie wavelength of the photoelectron
E	= $E_{h\nu} - E_o$, the photoelectron kinetic energy
h	= Planck's constant
m_e	= electronic rest mass

These photoelectrons may be scattered by neighboring atoms. The mechanism by which the scattering is expressed in the X-ray absorption spectrum is not transparent, but conceptually the EXAFS may be treated as an electron scattering pattern when plotted against wave vector K, where:

$$K = 2\pi/\lambda_e = 2\pi[2m\,(E_{h\nu} - E_o)]^{1/2}/h$$

A maximum in the EXAFS corresponds to the condition in which the distance between the absorbing atoms and the scattering neighbors is an integer multiple of the photoelectron wavelength, while a minimum reflects a half-integer separation. Of course, the experimental spectrum is a sum over all the absorbing sites and all the scattering neighbors. The EXAFS function is described by:

$$\chi(K) = \sum_i \sum_j \frac{N_{ij}}{K^2 R_j^2} F_{ij}(K) \exp(-2K^2 \sigma_{ij}^2) \sin[2KR_j + \phi_{ij}(K)]$$

where K is as defined above, and:

N_{ij} = number of atoms of atomic number i in shell j
R_j = radius of coordination shell j
$F_{ij}(K)$ = electron back-scattering function, dependent on the atomic number of the scatterer
σ_{ij}^2 = mean square displacement about interatomic distance R
$\phi_{ij}(K)$ = electron scattering phase shift function, dependent on atomic number of both absorber and scatterer.

The sums are taken over all the different atomic numbers represented among the neighboring atoms, and over all coordination shells. The EXAFS thus contains information on the identity, number, and distance of neighboring atoms to the absorber atoms, and the disorder of these atoms. In practice, the information in EXAFS is limited by several factors to radii of less than about 1 nm.

The situation is really quite analogous to X-ray diffraction, in which the scattered particle is a photon and the distance probed is the projection of interplanar spacing onto the X-ray trajectory, varied by rotating the sample with respect to the source. There are, however, some useful differences between EXAFS and XRD. First, the source of the scattered photoelectron in EXAFS is an atom of the chosen element, and only the vicinity of such atoms is illuminated, instead of the entire bulk phase, which may not be interesting. Second, the scattered photoelectron carries a signature from the scattering atom, which identifies its atomic number, in contrast to X-ray scattering amplitude, which is structureless. Third, EXAFS is a short-range phenomenon because the mean-free path of a photoelectron is short and is not dependent on crystallinity. Finally, the sample is stationary and the placement of detectors is flexible, which facilitates the design of in situ measurements.

7.3
Scope of Applicability of XAS

7.3.1
Applicability to Elements

X-ray absorption spectra are measurable from any of the elements. Experimental considerations, such as sample characteristics and accessible in situ conditions,

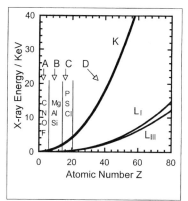

Fig. 7.4 Relationship between atomic number and absorption edge energies [11]. K-edges are stronger than L-edges and are used to measure spectra until their energy positions are prohibitively high, around Sn. Regions A–D present four distinct sets of experimental considerations for the measurement of spectra.

are determined mainly by the energy range of the spectrum, which is dictated by the atomic number of the element to be studied. Figure 7.4 shows a plot of atomic number vs. absorption edge energy [11]. Several regions in the plot presenting distinct experimental challenges are discussed below. In situ studies under catalytically relevant conditions are hardly practical with elements lighter than Ti, but the light elements are discussed briefly for completeness.

7.3.1.1 Low-Z elements (C, N, O, and F)

XAS is most difficult to apply to low-Z elements (region A in Fig. 7.4). The X-rays used to measure the spectrum are very low in energy, and consequently are strongly scattered and non-penetrating. Indeed, the same radiation in this energy regime is called X-ray when used to excite core levels of light atoms and vacuum ultraviolet (VUV) when used to excite valence levels of heavier atoms. Nevertheless, the technique has been successfully applied to light elements in catalysts [22–24]. For example, XAS has been used to characterize coke on a zeolite [22] and fluoride on alumina [23].

Special experimental modifications for XAS of low-Z elements include a vacuum ultraviolet source such as NSLS or ALS, a windowless vacuum beamline equipped with a scanning monochromator (grating-type dispersive element), a sample that is sufficiently thin or dilute to avoid distortion of the spectrum, and a detector sensitive in the low-energy range [24].

7.3.1.2 Mg, Al, and Si

The elements Mg, Al, and Si, with K absorption edges at 1303, 1559, and 1839 eV, respectively (9.5, 8.0, and 6.7 Å), region B in Fig. 7.4, present a unique XAS problem. The generation of monochromatic X-rays requires a dispersive element with a characteristic period similar to, but longer than, the desired wavelength. Unfortunately, the wavelength ranges required for the spectra of these elements are too short for ruled gratings, but longer than the unit cell dimensions of all but a few substances that can be grown as sufficiently large single crystals. Crystals for this range of wavelengths include InSb, topaz [$Al_2(F,OH)_2SiO_4$], quartz (SiO_2), and beryl ($3BeO \cdot Al_2O_3 \cdot 6SiO_2$). InSb is a malleable intermetallic phase that has relatively poor resolution and efficiency. The oxides contain Si or Al, which introduce unacceptable structure into the monochromatic beam in the energy range of interest. The oxides also have poor thermal and electrical properties and degrade rapidly [25]. The dispersive element of choice in the 1–2 KeV range is "YB_{66}", which actually has over 1600 atoms per unit cell and a lattice parameter of 23.44 Å. These crystals are currently available on SSRL Beamline 3–3 [25, 26].

Aside from the problem of preparing the necessary monochromatic X-rays, the X-rays are also soft and non-penetrating (Fig. 7.2), so that vacuum conditions and soft X-ray-sensitive detectors are still required as for the low-Z elements noted above.

7.3.1.3 P, S, and Cl

These elements are easily studied on many beamlines. The low energies of the K edges (2.1, 2.5, and 2.8 KeV for P, S, and Cl, respectively; region C in Fig. 7.4) require a monochromator with Si(111) crystals that can reach a large Bragg angle. Because of the low X-ray energies involved, a vacuum or He beam-path must be provided and spectra are always measured in fluorescence or electron-yield mode. Samples must be extremely thin or dilute to avoid distortion.

7.3.1.4 High-energy edges (Z > 21)

Spectra from elements heavier than Sc present no special problems (region D in Fig. 7.4). The K-edges present larger cross-sections than the L-edges, and are usually used up to Sn (29 KeV). Above 29 KeV, sources lose brightness, monochromatization and detection become problematic, and core-hole lifetime broadening degrades resolution to an unacceptable extent [27]. Therefore, the 5d elements Cs to Pb and the lanthanides are usually studied using L-edges, of which the L_{III} edge is the strongest.

7.3.2
Accessible In Situ Conditions

As noted above, elements lighter than Ti require that spectra be measured in vacuo or in a low pressure of He or H_2. The energies of the first-row transition metal

K-edges are high enough to overcome this obstacle, although in situ cell design must still incorporate thin windows of low-Z construction, minimize beam paths through gases at high pressure, and facilitate fluorescence detection since transmission measurements might not be practical due to strong scattering. Sample temperature is irrelevant to the mechanics of spectrum measurement, although the information content of EXAFS is decreased when the sample is hot. Electric fields, such as in an electrochemical cell, and illumination of the sample at some other wavelength, are irrelevant to the measurement of the spectrum.

Cell designs for in situ XAS have been reviewed [10]. The XAS technique probably imposes fewer constraints on cell design than any other spectroscopic technique. Perhaps the best in situ cell design for studying heterogeneous catalysts with gas-phase reactants was reported by Clausen et al. [28]. The catalyst is loaded into a 1 mm diameter quartz capillary tube, which serves as the window and as a tubular reactor with good hydrodynamics. The sample charge is less than about 50 mg, and typical gas flow rates of 1–5 $cm^3 min^{-1}$ present less of a safety issue than in other cell designs requiring higher flow rates. Cells for electrocatalysis [29], use under supercritical conditions [30], and simultaneous XRD and XAS have been described [28, 31, 32].

7.4
Mechanics of Measurement

This discussion is restricted to the measurement of spectra in the hard X-ray range (energy greater than 4 KeV), since in situ measurements are impractical with lower

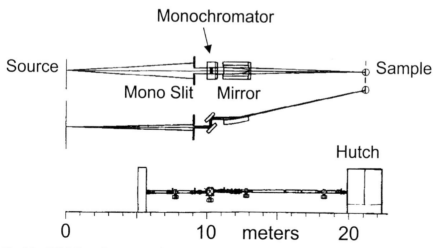

Fig. 7.5 NSLS Beamline X-10C. The beamline is a conventional spectrometer, albeit larger than usual. Since X-rays are difficult to focus and resolution is determined by beam divergence, distances are long. The mirror is not implemented on all beamlines.

energy X-rays. Laboratory X-ray absorption spectrometers have been reported [33,34] but have not enjoyed great success. The technique requires a synchrotron X-ray source for practical purposes. Links to many synchrotron X-ray sources worldwide are provided at [1].

7.4.1
The XAS Spectrometer – The Beamline

The XAS spectrometer is called a beamline (Fig. 7.5). Optical elements, besides the source, are a monochromator slit that usually controls resolution, a monochromator, and the detectors. A mirror for rejection of harmonics is provided on some beamlines. The detectors and the sample are housed inside a shielding enclosure called a hutch.

7.4.2
Detectors

7.4.2.1   Ion chambers

Ionization chambers are the most widely used detectors for XAS and the only type used for transmission measurements. They comprise two biased plates, between which the X-ray beam passes. A gas that is ionized by the beam floods the gap. A current flows between the plates in proportion to the beam intensity. Advantages include wide linear response range and high maximum count rate, low price, high durability, and sensitivity in different energy ranges selectable by choice of fill gas (heavier gas at higher energies). Many samples contain a low concentration of the absorber, so that fluorescence detection is preferred over transmission detection. Ion chambers that present a large window, for detection across a large solid angle, are widely used for measuring fluorescence spectra [35].

7.4.2.2   Solid-state detectors

The Ge and Si(Li) solid-state energy dispersive detectors are used when the absorber concentration is extremely low (less than about 1000 ppm). These detectors offer photon counting efficiencies near unity and allow energy discriminations, $E/\Delta E$, of about 100 or 50 for Ge and Si(Li) types, respectively. However, maximum count rates are low, so that the sensitivity advantage is lost if the sample is strongly scattering. The detector elements are small, e.g. 10 mm^2, and the detectors are fragile, expensive, and somewhat complex.

7.4.2.3 **Proportional counters**

Proportional counters provide less energy resolution than solid-state detectors and a narrow linear range, but may be preferred at low X-ray energies because they provide gain. Proportional counters for XAS have been discussed by Fischer [36].

7.4.2.4 **Electron yield detectors**

Electron yield detection is generally regarded as a high-vacuum technique, but in situ cells that incorporate electron detectors have been designed for use in the presence of high-pressure He or H_2 [37, 38]. The sample must be electrically conducting or else the method fails due to sample charging, but in favorable cases electron yield detection provides extremely high sensitivity. The method is somewhat surface sensitive, due to the short mean-free path of electrons throughout most of the kinetic energy range of EXAFS [39]. The signal originates within roughly 1000 Å of a surface, but this can be decreased by energy-filtering the electrons reaching the detector [40, 41]. Even when surface sensitivity is not in itself important, electron yield detection is useful for samples that cannot be prepared in thin or dilute form, since the signal originates within a sufficiently thin slab that distortions of spectra due to excessive sample thickness are avoided.

7.5
Limitations

Although laboratory X-ray absorption spectrometers have been reported, almost all spectra are measured using X-rays from a synchrotron radiation source. This is because the technique requires an intense, broad-spectrum X-ray source. The Bremstrahlung background beneath the emission lines from laboratory X-ray sources is isotropic and has insufficient intensity in a narrow energy bandpass, in a beam that can be focussed onto the sample. Laboratory spectra take days to measure, which tests the stability of both the sample and the optics. In contrast, synchrotron radiation is obtained in a narrow, well-collimated beam, with a continuous energy spectrum from infrared into the gamma region. At present, the US has four hard X-ray and five soft X-ray/VUV synchrotron sources. Each supports many experiments simultaneously. It is certainly a nuisance that one must travel to measure spectra. On the other hand, the sources and spectrometers (beam lines) are expensive, complex, and fragile and would not be pleasant additions to a laboratory! They are operated and maintained by the National Laboratories and there are no charges to use them for non-proprietary experiments. Proposals to obtain beam time are, however, highly competitive.

XAS also presents more fundamental limitations. Analysis of the XANES region of spectra remains empirical because calculation of XANES requires so much computing power. Analysis of EXAFS requires experience and a strong background in the materials science of the samples. The problem is that EXAFS yields a one-dimensional radial distribution function. The information content is sufficient to test a moderately complex structural model, but is not often sufficient to define an unknown structure. For example, the Fourier-transformed Mo K-edge EXAFS measured during in situ calcination of 4% Mo/H-ZSM5, and refined simulations generated by the FEFF6.0 program [42], are shown in Fig. 7.6. The oxide simulation (Fig. 7.6a) involved six independent single-scattering contributions from the

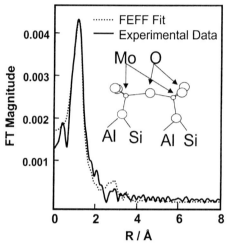

Fig. 7.6a Solid trace: Fourier-transformed Mo K-edge EXAFS from precursor of 4% Mo/H-ZSM5 catalyst for non-oixidative methane aromatization (20% O_2/He 973 K) [19]. TPR and in situ Raman and NMR spectroscopies suggested the structure, which was confirmed by simulation of the EXAFS [18, 43]. Broken trace: Refined FEFF6.0 simulation indicating that the EXAFS is consistent with the Mo dimer shown in the inset.

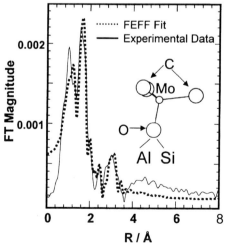

Fig. 7.6b Solid trace: Fourier-transformed Mo K-edge EXAFS magnitude from carburized 4% Mo/H-ZSM5. The fit (broken trace) shows three carbon neighbors at 2.09 Å, one oxygen neighbor at 1.71 Å, and two Al and/or Si neighbors at 3.7 Å. The average number of Mo neighbors within 3 Å is only about 0.1. The Mo–C distance is close to the first-shell Mo–C distance in Mo_2C, but the bulk structure is not formed.

first and second coordination shells of the Mo atoms, and 20 dependent multiple-scattering contributions. The second-shell Mo, Al, and Si neighbor contributions interfere destructively among themselves and as a result are hardly visible. The variables were controlled by constraining multiple scattering shells to conform to single-scattering shells, and by constraining single-scattering parameter values within the physically plausible range. Acceptable parameter values were expected to compare closely to bond lengths and site structures previously reported for Mo compounds.

7.6
Examples of Applications

7.6.1
Mo/H-ZSM5 Catalyst for Non-oxidative CH_4 Reactions

Catalysts comprising Mo supported on H-ZSM5 have been studied by Borry et al. [19, 43] for non-oxidative reactions of CH_4. The solid-state exchange of MoO_3 into H-ZSM5 by heating the mixture in air was monitored by in situ XANES (Fig. 7.3) and Raman spectroscopy, demonstrating the spreading of the MoO_3 by delamination onto the zeolite, and then dispersal within the zeolite channels as a dimeric species resembling the $MgMo_2O_7$ oxyanion. The $Mo_2O_7^{2-}$ dimeric Mo structure was simulated using the program FEFF6.0 [42]. The simulation, which involved six single and 20 multiple-scattering shells, was refined in real space to the experimental EXAFS. The resulting fit is shown in Fig. 7.6a. It must be emphasized that the spectroscopic data were sufficient to test a complex model, but could not have provided a structural determination in the absence of the model. The exchanged Mo/H-ZSM5 containing the dispersed Mo species was then treated with CH_4 and the evolution of the carburized Mo species was observed by EXAFS while the induction of activity was monitored independently. The Mo active site incorporated three carbon neighbors at distances similar to the Mo–C distance in Mo_2C, but without second-shell Mo neighbors, reflecting the high dispersion of the Mo in the active catalyst [19]. The corresponding fit to the EXAFS Fourier-transform magnitude is shown in Fig. 7.6b.

7.6.2
Cu/ZnO Methanol Synthesis Catalyst

A result from a study of Cu-based methanol synthesis catalysts by the Haldor Topsøe group, employing a capillary in situ cell, is shown in Fig. 7.7. A comparison was made between 5% Cu/SiO_2 and 5% Cu/ZnO catalysts, which were reduced in $CO/CO_2/H_2/Ar$ (0.5:4:4:91.5) mixtures at 493 K and then used in methanol synthesis

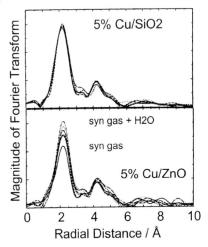

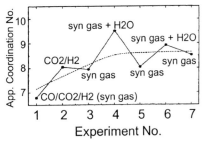

Fig. 7.7 Reconstruction of Cu metal function in Cu/ZnO methanol synthesis catalyst in response to changes in redox potential of feed, Cu particles flattening in response to more reducing potential in feed [28, 44–46].
Left: The magnitude of the peak at 2 Å reflects the number of Cu atoms in the first Cu coordination sphere. Cu particles on SiO_2 were insensitive to the presence of H_2O in the feed, but agglomerated on ZnO in the presence of H_2O.
Right: Cu–Cu coordination numbers calculated from the EXAFS of Cu/ZnO are plotted versus the number of the wet/dry cycle.

with either CO_2/H_2 (10:90), $CO/CO_2/H_2$ (5:5:90), or $H_2O/CO/CO_2/H_2$ (3:4.85:4.85:87.3) mixtures, all at atmospheric pressure [28, 44]. In both catalysts, with all synthesis gas compositions, Cu(0) was the most abundant species at 493 K. The structures and coordination numbers of Cu species on the SiO_2 support were found to be insensitive to the composition of the gas phase. In contrast, the Cu–Cu coordination number for Cu metal particles on ZnO changed reversibly as the feed composition was varied (Fig. 7.7). The Cu particles became more dispersed (i.e. became smaller) when water was removed from the synthesis gas stream. These data led to a model that correlated the relative interfacial areas between Cu metal and ZnO in Cu/ZnO with the reduction potential of the gas mixture, which determined the extent of reduction of the ZnO component and the consequent spreading of Cu metal species on ZnO [45]. A rate law was developed that incorporated H_2O partial pressure only in a term that predicted its effect on Cu surface area. This successfully predicted rates measured under conditions differing only in H_2O concentration. The authors concluded that reconstruction of Cu particles in the presence of H_2O provided a more reasonable explanation than competitive adsorption for the observed inhibition of methanol synthesis reaction rates by H_2O [46].

References

1 The web site http://www.wmich.edu/~physics/research/photon/sources.html provides links to 21 synchrotron sources around the world.

2 R. Stumm von Bordwehr, *Ann. Phys. Fr.* **14**, 377 (1989).

3 P. H. Lewis, *J. Phys. Chem.* **64**, 1103 (1960); **66**, 105 (1962); **67**, 2151 (1963).

4 R. A. Van Nordstrand, in *"Röntgenspektren und Chemische Bindung, Vortrage des Internationalen Symposiums, 23–24 September 1965*, Leipzig, Physikalisch-Chemisches Institut der Karl Marx Universität Leipzig (1966), pp. 255–267.

5 F. W. Lytle, G. H. Via, J. H. Sinfelt, *J. Chem. Phys.* **67**, 3831 (1977).

6 P. A. Lee, P. H. Citrin, P. Eisenberger, B. M. Kincaid, *Rev. Mod. Phys.* **53**, 769 (1981).

7 M. J. Fay, A. Proctor, D. P. Hoffmann, D. M. Hercules, *Anal. Chem.* **60**, 1225A (1988).

8 J. J. Rehr, A. Ankudinov, S. I. Zabinsky, *Catal. Today* **39**, 263 (1998).

9 G. Meitzner, *Catal. Today* **39**, 281 (1998).

10 G. Meitzner, S. R. Bare, D. Parker, H. Woo, D. A. Fischer, *Rev. Sci. Instrum.* **69**, 2618 (1998).

11 J. A. Bearden, A. F. Burr, *Rev. Mod. Phys.* **39**, 125 (1967).

12 W. H. McMaster, N. Kerr Del Grande, J. H. Mallett, J. H. Hubbell, Lawrence Livermore National Laboratory Report UCRL-50174 Section II Rev. I (1969). Available at http://cars.uchicago.edu/~newville/mcbook/

13 D. Calloway, *J. Chem. Educ.* **74**, 744 (1997).

14 V. Kunzl, *Coll. Czech. Chem. Commun.* **4**, 213 (1932).

15 N. E. Pingitore, Jr., G. Meitzner, K. M. Love, *Geochim. et Cosmochim. Acta* **59**, 2477 (1995).

16 J. H. Sinfelt, G. D. Meitzner, *Acc. Chem. Res.* **26**, 1 (1993).

17 F. Farges, G. E. Brown, Jr., J. J. Rehr, *Geochim. et Cosmochim. Acta* **60**, 3023 (1996).

18 G. Meitzner, E. S. Huang, *Fresenius J. Anal. Chem.*, **342**, 61 (1992).

19 W. Li, G. Meitzner, R. W. Borry III, E. Iglesia, *J. Catal.* **191**, 373 (2000).

20 S. R. Wasserman, *J. Phys. IV Fr.* **7**, C200–203 (1995).

21 M. Fernandez-Garcia, C. Marquez-Alvarez, G. L. Haller, *J. Phys. Chem.* **99**, 2565 (1995).

22 S. M. Davis, Y. Zhou, M. A. Freeman, D. A. Fischer, G. M. Meitzner, J. L. Gland, *J. Catal.* **139**, 322 (1992).

23 S. M. Davis, G. D. Meitzner, D. A. Fischer, J. Gland, *J. Catal.* **142**, 368 (1993).

24 D. A. Fischer, G. Meitzner, J. Gland, in *"X-ray Absorption Fine Structure"* (Ed.: S. Hasnain), Ellis Horwood Ltd., New York, 1991, pp. 619.

25 J. Wong, G. N. George, I. J. Pickering, Z. Rek, M. Rowen, T. Tanaka, G. H. Via, B. DeVries, D. E. W. Vaughan, G. E. Brown, Jr., *Solid State Commun.*, **92**, 559 (1994).

26 M. Rowen, Z. U. Rek, J. Wong, T. Tanaka, G. N. George, I. J. Pickering, G. H. Via, G. E. Brown, Jr., *J. Synchrotron Rad.* **6**, 25 (1993).

27 M. O. Krause, J. H. Oliver, *Phys. Chem. Ref. Data* **8**, 329 (1979).

28 B. S. Clausen, B. Steffensen, B. Fabius, J. Villadsen, R. Feidenhans'l, H. Topsøe, *J. Catal.* **132**, 524 (1991).

29 D. A. Scherson, S. Sarangapani, F. L. Urbach, *Anal. Chem.* **57**, 1501 (1985).

30 D. M. Pfund, J. G. Darab, J. L. Fulton, Y. Ma, *J. Phys. Chem.* **98**, 13102 (1994).

31 M. G. Samant, G. Bergeret, G. Meitzner, M. Boudart, *J. Phys. Chem.* **92**, 3542 (1988).

32 A. J. Dent, M. Oversluizen, G. N. Greaves, M. A. Roberts, G. Sankar, C. R. A. Catlow, J. M. Thomas, *Physica B* **208**, 253 (1995).

33 Y. Udagawa, *The Rigaku Journal* **6**, 20 (1989).

34 D. C. Koningsberger, in *"X-ray Absorption: Principles, Applications, Techniques of EXAFS, SEXAFS, and XANES"* (Eds.: D. C. Koningsberger, R. Prins), John Wiley & Sons, New York, 1988, pp. 163.

35 E. A. Stern, S. M. Heald, *Rev. Sci. Instrum.* **50**, 1579 (1979).

36 D. A. Fischer, J. Colbert, J. L. Gland, *Rev. Sci. Instrum.* **60**, 1596 (1989).

37 M. E. Kordesch, R. W. Hoffman, *Phys. Rev. B* **29**, 491 (1984).

38 F. W. Lytle, R. B. Greegor, G. H. Via, J. M. Brown, G. Meitzner, *J. de Physique, Colloque* **C8**, suppl. 12, C8–149 (1986).

39 W. E. Spicer, in *"Optical Properties of Solids: New Developments"* (Ed.: B. O. Seraphin), North Holland, Amsterdam, 1976, pp. 631.

40 Zw. Bonchev, A. Jordanov, A. Minkova, *Nucl. Instrum. Meth.* **70**, 36 (1969).

41 *"Workshop on Scientific Directions at the Advanced Light Source: Summary and Reports of the Working Groups;* E. O. Lawrence Berkeley National Laboratory, 23–25 March 1998. Available from Mat. Tech. Inf. Serv., U.S. Dept. Commerce, July 1998, p. 141.

42 S. I. Zabinsky, J. J. Rehr, A. Ankudinov, R. C. Albers, M. J. Eller, *Phys. Rev. B* **52**, 2995 (1995).

43 R. W. Borry, Y.-H. Kim, A. Huffsmith, J. A. Reimer, E. Iglesia, *J. Phys. Chem.* **103**, 5787 (1999).

44 B. S. Clausen, L. Grabaek, G. Steffensen, P. L. Hansen, H. Topsøe, *Catal. Lett.* **20**, 23 (1993).

45 B. S. Clausen, J. Schiotz, L. Grabaek, C. V. Ovesen, K. W. Jacobsen, J. K. Norskov, H. Topsøe, *Topics Catal.* **1**, 367 (1994).

46 C. V. Ovesen, B. S. Clausen, J. Schiotz, P. Stoltze, H. Topsøe, J. K. Norskov, *J. Catal.* **168**, 133 (1997).

8

In Situ Measurement of Heterogeneous Catalytic Reactor Phenomena using Positron Emission

B. G. Anderson[1], A. M. de Jong[2] and R. A. van Santen[1]

8.1
Introduction

Chemists and chemical engineers studying the kinetics of heterogeneously-catalysed reactions are constantly trying to answer questions regarding the chemical nature of reacting species and the rates at which these species undergo transformations. Techniques are now available that are capable of the in situ measurement of chemical species in a quantitative manner. Many spectroscopic and surface science techniques have been successfully applied to this problem over the past thirty years and have generated a large body of data. However, the experimental conditions under which these measurements must be performed (such as the temperature and pressure) often differ considerably from those that exist under normal process conditions. Thus, a question remains as to whether the data obtained can be extrapolated to the real system. As these data have generally been obtained under conditions differing from those of the actual process, a link must be made back to the original systems of interest occurring in chemical reactors at elevated temperatures and at atmospheric pressure and above. Mathematical models based on previously determined mechanistic information provide this link.

Mathematical models of reactor kinetics attempt to describe the concentration distributions of various reactants, intermediates, and products within the reactor during the course of the reaction. In order to validate these models, concentration distributions must be measured. However, as these chemical reactions under real process conditions generally occur within metal reactors enclosed by a heating mantle, it is not a simple matter to "look inside". Rather, most techniques rely on measurements of concentration distributions measured at the reactor inlet and outlet only.

Ideally, one would like to have a technique that could "image" the reactor in such a way as to be able to determine both the chemical identity and quantity of all reactants, intermediates, and products as a function of time and position under identical conditions of temperature, pressure, flow rate, etc. In addition, it is desirable that the probe method does not disturb the reaction, that is the method should be non-invasive.

Medical research has provided an incentive to develop several sophisticated non-invasive, in situ, imaging techniques. Techniques such as magnetic resonance imaging (MRI) and computer-assisted tomography (CAT) are now widely used diagnostic tools to study structure within the living human. Other techniques based on the detection of emitted radiation from injected radioactive tracers are used to study biochemical and metabolic functions.

These techniques have not as yet been widely applied in engineering and catalysis research, although interest is growing. Industrial applications of nuclear magnetic resonance have recently been reviewed by Gladden [1]. An emerging field is that of "Process Tomography" [2, 3]. In this chapter, we discuss recent adaptations and applications of one specific type of imaging technique, Positron Emission Tomography (PET), to catalysis research. Although PET is now well established as a diagnostic tool for in vivo imaging of the function of human organs, particularly of the brain and the heart [4], it is probably unknown to most researchers outside of the medical community. Thus, we start with a description of the technique and the principles upon which it is based.

8.1.1
Positron Emission and Positron-Electron Annihilation

The decay of radioactive isotopes through electron emission, so-called beta decay, is well known to most scientists. In this mode, unstable nuclei that have an excessive number of neutrons, for example ^{14}C, can emit fast electrons, β^- particles, in order to attain a stable nuclear configuration. Nuclei with insufficient neutrons, such as ^{11}C, can obtain stability by emitting fast positrons, β^+ particles (the anti-matter equivalents of electrons). Both processes are classified as radioactive β decay. In each case, the mass number of the nucleus remains constant but the atomic number changes.

Positron-emitting isotopes of many different elements exist, including: ^{11}C (half-life ($t_{1/2}$) = 20.4 minutes), ^{13}N ($t_{1/2}$ = 9.96 minutes), ^{15}O ($t_{1/2}$ = 2.07 minutes), and ^{18}F ($t_{1/2}$ = 109.8 minutes). Since all of these radioisotopes are short-lived they must be produced on-site. This is normally accomplished by irradiation of an appropriate target material with energetic beams of protons or deuterons supplied by a cyclotron.

Since the positron is the antiparticle of the electron, an encounter between them can lead to their subsequent mutual annihilation. Their combined rest mass energy then appears as electromagnetic radiation. Annihilation can occur by several mechanisms: direct transformation into one, two, or three photons; or the formation of an intermediate, hydrogen-like bound state between the positron and the electron, called a positronium atom (Ps). The extent to which each annihilation mechanism contributes depends on the kinetic energy of the positron-electron pair.

Positrons emitted during the β^+ decay process possess a statistical distribution of kinetic energies ranging from zero to a maximum value, T_{max}, which is dependent on the decaying nucleus ($T_{max} = 0.96$ MeV for ^{11}C). The average kinetic energy is equal to 0.4 T_{max}. The probability of annihilation is negligibly small at high energies. The emitted positrons must therefore be slowed down by inelastic scattering interactions between the nuclei and the bound electrons within the surrounding medium to near thermal values before annihilation can occur. The lifetime of a positron is of the order of tens of nanoseconds. During its lifetime, the positron will travel a distance, known as the stopping distance, which is dependent on its energy and on the density of the surrounding material. For 0.4 MeV positrons (average kinetic energy of positrons emitted from ^{11}C) in a medium with a density of 0.5 g cm^{-3} (such as a zeolite or metal oxide), this corresponds to ca. 3 mm [5].

The predominant annihilation process for thermalized positrons is by the direct production of two photons. If both the positron and the electron were at rest upon annihilation, conservation of energy dictates that each emitted photon would have an energy equal to the 511 keV rest mass energy of the positron or electron. Conservation of momentum would dictate that the two gamma photons be emitted in opposite directions to one another, since the initial momentum of the positron-electron pair was zero.

However, even when the kinetic energy of the positron is zero upon annihilation, the average kinetic energy of the electrons in the surrounding medium is non-zero, typically ca. 10 eV [6]. Thus, the positron-electron pair will, on average, have a non-zero momentum. The transverse component of the momentum gives rise to a small deviation from the 180° emission angle (full-width-at-half-maximum of the angular spread ca. 0.4°) and to a small energy shift [7]. The longitudinal component of the momentum gives rise to a slight Doppler shift. The maximum energy shift is 1.9 keV [8]. The contribution of other decay mechanisms is rather small when compared with that of two-photon decay; 97% of the positrons emitted by ^{11}C nuclei are first thermalized and then annihilate to form two gamma photons possessing energies of 511 ± 2 keV. The emission angle between these photons is 180° ± 0.4°. The remainder of the positrons are either annihilated in-flight to form two photons (2%) or through three-photon emission (1%) [9].

8.1.2
Detection Methods Based on Positron Emission

This chapter focuses on the use of molecules labeled with positron-emitting nuclei and imaging techniques based on their detection. We therefore only briefly mention the recent use of positrons as analytical probes, so-called Positron Annihilation Spectroscopy (PAS), to study the structure of heterogeneous catalysts. In PAS, a sample is irradiated with either mono- or poly-energetic beams of positrons. By measuring quantities such as the lifetime of the positrons, the energy distribution of the scattered positrons, or their angular distribution, it is possible to study the electronic properties of the surface or bulk of a solid. The interested reader is referred to recent reviews on this subject [10–12].

The emitted gamma photons produced by positron-electron annihilation can be detected using scintillation crystal detectors such as sodium iodide (NaI) or bismuth germanium oxide (BGO). The short half-life leads to high specific activity. Only a very small quantity of radiolabeled molecules is thus required, making positron techniques very non-invasive. In fact, practical catalyst studies can be carried out using less than 37 kBq of carbon-11. This corresponds to less than 6.5 ± 10^7 molecules or 6.5 ± 10^{-8} of a monolayer coverage for a 1 cm^2 single-crystal surface.

The first use of positron-emitting isotopes as tracers in catalysis research was published in 1984 by Ferrieri and Wolf [13, 14]. ^{11}C-labeled acetylene and propylene were employed to monitor the alkyne cyclotrimerization reaction on silica/alumina-supported chromium(VI). Single annihilation photons were collimated and detected using a sodium iodide scintillation detector. The aromatic products (benzene, toluene, xylenes) desorbed from the surface were analyzed using radio gas chromatography. These authors named their technique Positron Annihilation Surface Detection (PASD).

More recently, Baiker and co-workers [15] used ^{13}N-labeled NO to investigate the selective catalytic reduction (SCR) of NO by NH$_3$ over vanadia/titania at very low reactant concentrations. Concentrations of 5 ± 10^{-9} ppm ^{13}NO were used (10^{11} times lower than those typically used). Again, only single annihilation photons were detected by NaI scintillation detectors placed after the reactor bed.

A second detection method involves the coincident detection of both of the photons produced by the annihilation event. This can be achieved by using two scintillation detectors, each placed on opposite sides of the emitting source. In this mode, only pairs of detected events that occur within a preset coincidence window (typically less than 50 ns) are counted. The position of the annihilation event that gave rise to the two detected photons can then be located along a chord joining the two detector elements. The concentration of the radiolabeled isotope at that position can also be determined by integrating the number of events detected during a fixed time.

Due to the penetrating power of the emitted 511 keV gamma photons, which can pass through several millimeters of stainless steel, detection is possible from within steel reactors or process vessels. The coincident detection of photons is the principle of techniques such as Positron Emission Tomography (PET), Positron Emission Particle Tracking (PEPT), and Positron Emission Profiling (PEP), which are discussed below.

8.1.3
Positron Emission Tomography (PET), Particle Tracking (PEPT) and Profiling (PEP)

Positron Emission Tomography (PET) was developed over 20 years ago and is now well established as a diagnostic technique in nuclear medicine, providing 3D images of the distribution of radiolabeled molecules within living human organs. The development of a new breed of small self-shielding cyclotrons in the 1980s and significant improvements in computer hardware and software has led to an explosive growth in the numbers of PET facilities worldwide. It has recently been predicted that PET will soon become standard in all nuclear medicine facilities [16]. Unlike CAT and MRI, which measure structural information, PET is capable of providing rate information regarding biochemical and metabolic processes. Injected radiotracers such as [^{18}F]2-fluoro-2-deoxyglucose and [^{13}N]ammonia are used to make in vivo measurements of cerebral metabolism and myocardial blood flow. Feliu has published an excellent overview of this technique [17].

Medical PET detectors normally employ one or more rings of small scintillator detectors. The NeuroECAT tomograph, for example, consists of eight detector banks arranged in an octagonal pattern. Each bank contains 11 BGO scintillation detectors. Individual detector elements in opposing banks form coincidence pairs. During a scan, the tomograph is rotated about an axis parallel to the subject and is linearly translated along the same axis. In this manner, photons emitted over 360° in the plane of the detectors can be recorded. Using tomographic reconstruction techniques, the data can be used to map the distribution of the positron emitter in a slice through the subject. Time is required for rotation and translation of the tomograph and to acquire sufficient coincident events for adequate measurement statistics. As a result, PET spectra generally require scan times of the order of 10 to 15 minutes, during which time millions of coincidence events will be collected by the scintillator elements, thus enabling a full 3D reconstruction of the imaged object to be produced. As pointed out above, the annihilation position is somewhat different from the original position of radioactive decay since the positron must first be slowed down. This stopping distance determines to a large extent the maximum attainable spatial resolution of the technique. Spatial resolutions of 8.5 mm × 8.5 mm in-plane and 13.5 mm in the axial direction can typically be achieved with the NeuroECAT tomograph. Current "state-of-the-art" PET detector technology based on BGO block-detectors is capable of achieving an intrinsic spatial resolution of between 3 and 4 mm in each direction [18]. Significant advances in small-scale 3D PET detectors suitable for applications such as imaging of small-scale anatomies of laboratory animals during drug trials have recently been reported [19,20]. For example, a so-called MicroPET camera has recently been developed that is capable of resolving structures as small as a sesame seed, ca. 6 mm^3. This represents a tenfold improvement in resolution over conventional scanners [21].

PET has been applied to problems of industrial interest only recently [22]. PET has been shown to be capable of monitoring turbulent two-phase (liquid/gas) flows using injected solutions of aqueous $Na^{18}F$ as radiotracers [23].

Many dynamic processes occurring in industrial equipment, for example mixing processes, occur on time-scales that are too short for complete measurement of a 3D image. However, flow pattern measurements of such rapid processes can be measured if one restricts the measurements to the tracking of a single, radiolabeled, tracer particle. This technique, known as Positron Emission Particle Tracking (PEPT) [22, 24], has been used to measure physical processes such as powder mixing in a ploughshare batch mixer [25] and particle motion within a fluidized bed [26]. Both PET and PEPT have been used in a study of axial diffusion of particles in rotating drums [27]. Further studies on this subject proved to be very successful; slip of the bed at the walls was observed and axial dispersion coefficients were determined [28]. An improved PEPT system has recently been developed that is capable of continuously following the 3D trajectory of a radiotracer particle (as small as 500 µm) moving at 0.1 m s^{-1} with a resolution of 5 mm. The system has been used to measure in situ flow patterns of solids in a gas-solid Interconnected Fluidised Bed reactor [29].

Jonkers and co-workers conducted the first study in which PET was applied to chemical reactions in reactors [30, 31]. These pilot experiments were performed at the State University of Gent, Belgium, using a commercial PET camera designed for medical imaging, the NeuroECAT tomograph described above. The objective was to show that PET could be used to obtain images of a reaction occurring within a tubular, plug-flow reactor operating under normal process conditions, and that the data obtained could provide information for subsequent modelling of the reaction kinetics. These experiments will be discussed in more detail further on in this chapter.

Since the early 1990s, a facility has been developed at the Eindhoven University of Technology (TU/e) dedicated to the application of positron emission imaging for the study of physical and chemical processes that occur within a catalytic reactor under practical operating conditions. As measurements of concentration profiles in a single dimension are sufficient under axially-dispersed plug-flow conditions (since concentration gradients in the radial direction are negligible), a positron emission detector has been developed that is specifically tailored to the measurement of activity distributions as a function of time along a single, axial direction. This detector [32] is called a Positron Emission Profiling (PEP) detector to distinguish it from its 3D parent. The project is a collaborative effort between the Faculties of Chemical Engineering and Technical Physics in Eindhoven and the Department of Chemical Engineering at Delft University of Technology. The university's 30 MeV AVF cyclotron is used to prepare the positron-emitting nuclides ^{11}C, ^{13}N, and ^{15}O.

8.2
PEP Detectors and the Synthesis of Labeled Molecules Containing Positron-Emitting Isotopes

The following section describes the PEP detectors that have been designed and built at the Eindhoven University of Technology. Their basic operating principles and characteristics are highlighted.

In addition, the synthetic methods developed to produce several labeled molecules containing the positron-emitting isotopes, including ^{11}C, ^{13}N, and ^{15}O, are described.

8.2.1
The TU/e PEP Detector

The Positron Emission Profiling (PEP) detector is shown in Fig. 8.1. It has been designed to be flexible so that it may be used with a variety of different sizes of reactors; measurements can be carried out on reactors having bed lengths between 4.0 cm and 50 cm and diameters of up to 25 cm. The detector consists of two arrays of nine independent detection elements each, and is mounted horizontally, with

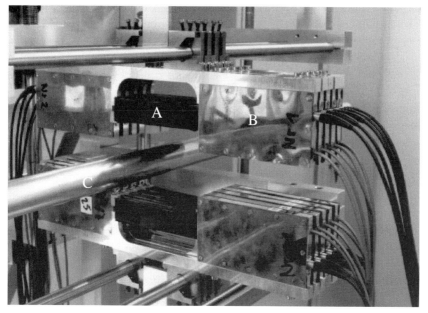

Fig. 8.1 A photo of the Positron Emission Profiling (PEP) detector showing the two banks of detector elements. Each element, a BGO crystal (A) is coupled to photomultiplier (B). An aluminum tube (C) is shown in place of the reactor.

the reactor and furnace placed between the upper and lower arrays. Each detection element consists of a bismuth germanium oxide (BGO) scintillation crystal coupled to a photomultiplier. The eighteen detection elements are situated in a frame, which allows adjustment of the overall detector dimensions if required.

The detection principle is as follows. Each detector element in a bank can form a detection pair with any of the elements in the opposite bank; the two banks of nine elements therefore form 81 (9^2) possible detection pairs. Coincident detection of the two 511 keV gamma photons, formed during positron-electron annihilation, by a detection pair is used to locate the position of the event along the cylindrical axis of the bed at the point at which the chord joining the two elements intersects this axis. Due to redundancy, the set of chords joining these 81 detection pairs results in 17 [(2×9) − 1] unique, equidistant positions along the cylindrical axis. A schematic diagram of the PEP detection principle is shown in Fig. 8.2. As stated above, each of the detector elements forms a possible detection pair with each of the nine elements in the opposite bank, thus forming 81 detection pairs. For clarity, a lower reconstruction mode is shown in the figure. In this mode, only coincident events detected in opposite and opposite-adjacent elements are used in the reconstruction. The coincidence time window is normally set to 40 ns since the time resolution, τ, has been determined to be 19 ± 1 ns [9] and 2τ is generally considered to be an appropriate criterion for determining coincident events. It should be pointed out here that the lifetime of an emitted positron is dependent upon the surrounding medium (the basis for Positron Annihilation Lifetime (PAL) spectroscopy mentioned above, see [10–12]). The lifetime of a positron in a solid is dependent on the electron density of the solid; for example, in an oxide-supported or zeolite catalyst the lifetime of a positron is of the order of 2 ns [9].

Temporal information is obtained by collecting data during a fixed sampling period. A minimum integration time of 0.5 s is required to obtain sufficient coincident events for reliable statistics and for on-line computation; thus, the temporal resolution is 0.5 s [32]. When the detector block is in its "close-packed configuration", that is when all of the detector elements are placed tightly together, the spatial resolution of the detector is 2.9 mm [9], which is comparable to state-of-the-art 3D PET detectors and is near the theoretical limit of the spatial resolution governed by the range of the positrons under study (the stopping distance).

In order to reduce errors resulting from the detection of Compton scattered photons (the contribution from which can be as high as 35% [9]), which lead to anomalous positioning of annihilation events, energy selection of the photons is also employed. Coincident events having a measured energy below the lower limit of a window centred at 511 keV are rejected as being due to scattered photons. Since the energy resolution of BGO scintillators is poor (only 28% FWHM), this window is quite large (350–700 keV). The angular resolution was found experimentally to be ±0.5° [5]. As a result of the above experimental limitations of the detector, effects

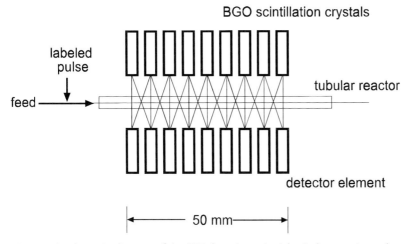

Fig. 8.2 A schematic diagram of the PEP detection principle. Only opposite and opposite adjacent reconstructions are shown for clarity.

such as emission angle deviation and Doppler energy shifts described above, have no discernible impact on the PEP images measured.

In summary, the above detection principle can be employed to obtain profiles of the average concentration of all radiolabeled components contained within volume segments of lengths greater than or equal to 3 mm in the imaged specimen at time intervals of 0.5 seconds.

Several of the applications of PEP described below were performed with the PEP detector described above. Recently, a new PEP detector has been developed at Eindhoven University of Technology. Better performance and increasing flexibility was the motive for its development. The new detector enables experiments in which larger catalyst beds are studied at high resolution, and offers higher count rates and online monitoring and processing of the data. The main features of this new detector will be briefly described in the next section. Further details of this system will be published soon.

8.2.2
The Improved PEP Detector

A new data acquisition system has been designed and is currently being implemented, which greatly enhances the performance of the PEP detector. This new detector system makes it possible to measure at higher count rates, resulting in faster measurements (better time resolution in PEP experiments) and with a higher dynamic range (improved sensitivity). The dynamic range becomes important for experiments that are performed over a longer period of time, in which a high activity is initially present in the reactor and thus high count rates are necessary. At the end of the experiment, the count rate will be low due to decay of the tracer compound. Also, when large differences in the concentration of the radiochemicals over the reactor bed arise, the dynamic range becomes important. Furthermore, the new detector allows for real-time acquisition and display of the concentration profiles. This greatly improves the feedback to the experimenter compared to a system that uses off-line reconstruction of the measured concentration profiles.

As mentioned above, Compton scattered photons lead to distortion of the measured profiles. Energy selection of the detected gamma photons reduces this effect. However, the poor energy resolution of the BGO detection elements (28%) causes a large overlap of the measured energy of scattered and non-scattered gamma photons. The new detector is equipped with the Dual Energy Window scatter correction technique [33, 34]. Gamma photons are now also measured in a second energy window in the Compton background. This technique assumes that the number of unscattered events in the photopeak window is a linear function of the number of events in both energy windows.

$$N_{photopeak,unscattered} = a \cdot N_{Compton} + b \cdot N_{photopeak}$$

This technique provides a method for correcting for adverse effects of the scatter of the annihilation radiation, which otherwise smear out the measured concentration profiles and affect accurate determination of catalytic properties.

At high count rates, the possibility exists that two gamma photons might be detected as coincident that do not arise from the same positron annihilation. These random coincidences give rise to incorrect position reconstructions. Since the number of random coincidences is proportional to the count rate, correction for their contribution is possible when the count rate of the individual detection elements is known. To correct for this phenomenon, nearly all detector systems are also programmed with a minimum time separation between two detected events. Below this interval, the two events are deemed non-coincident. This time is called the dead time, during which all other incoming events are lost. At high count rates, this loss of events becomes significant compared to the number of measured events, resulting in a nonlinear response. The dead time is also proportional to the number of detected events. The new detector provides the single count rates, not only for the dead time corrections, but also for the chance coincidence corrections.

Figure 8.3 presents an overview of the layout of all hardware components and of the flow of data and control of the PEP detector.

The heart of the new PEP detector consists of two parallel banks of 16 gamma-ray detection elements. The pre-amplifiers (PREAMP) convert the charge signal of the photomultiplier tubes into an energy signal and a timing signal. The energy

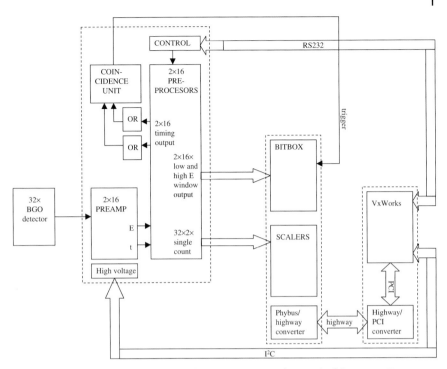

Fig. 8.3 A schematic overview of the data acquisition and control of the new PEP detector set-up.

signal is used for energy selection by the pre-processors (PREPROCESSOR). In the pre-processor, the energy of the detected gamma photon is determined, and whether this energy falls in the photopeak or Compton window. The timing signal is used by a coincidence circuit to detect whether a gamma photon was detected co-incidentally in the upper and lower detection element. The coincidence circuit consists of two 16-fold OR circuits (OR) for the two detector banks, the two outputs of which are processed by an AND circuit (AND).

The output of the coincidence unit is used to trigger the bitbox (BITBOX). This bitbox records the status of all pre-processors and reconstructs the annihilation position. By making a position histogram of a large number of these coincident events, a concentration profile is determined. This number of events is collected during each binning time, which can be programmed to be 0.1 s or longer. The number of events per binning time per detection element is also recorded. Two scalers perform this task (SCALER). These values are used to correct for dead time and chance coincidences.

To measure the count rate in the photopeak window and the total count rate, the pre-processor gives two types of scaler pulses. By counting the number of these scaler pulses per binning time, the single count rate of the pre-processors can be measured. One scaler pulse is given for each event that was processed by the pre-processor, and this pulse is used for dead-time corrections. When the energy of the gamma ray falls in the photopeak window another scaler pulse is given, which is used for chance coincidence corrections.

The data acquisition is controlled by a server PC running under the Vx/Works operating system, chosen because of its real-time capabilities. The data that are collected by the Vx/Works system are sent to the user interface, which is run on a separate PC. All functionality/control of the hardware through RS232, I2C, and PCI ports is implemented on the server.

8.2.3
Synthesis of Radiolabeled Molecules

The on-site preparation of radioisotopes is normally accomplished by irradiating an appropriate target material with energetic beams of protons or deuterons supplied by a cyclotron. During the irradiation process, the highly energetic particles impart large kinetic energies to the target molecules; these energies greatly exceed the bond dissociation energies. As a result, only very simple molecules survive as products. To produce more complex radiolabeled molecules containing positron emitters, post-irradiation chemical synthesis must be carried out to incorporate the smaller radiolabeled fragments into the larger framework. Strategies must be developed whereby the desired molecule can be produced and separated from other reaction products within a few half-lives of the radioisotope. This normally means that one must use a precursor that is only one or perhaps two reaction steps from the desired target molecule [17]. This requirement has not prohibited the synthesis of radiolabeled analogues of complex molecules such as synthetic drugs and drug metabolites [35]. Nor has the need for a dedicated cyclotron curtailed the use of this technique. In fact, it has led to the development of small, shielded, semi-automated cyclotrons, which are now commercially available [17, 36]. A number of labeled molecules containing ^{11}C, ^{13}N, or ^{15}O have been produced at the TU/e for use in PEP experiments on catalytic reactions. The production and synthesis methods of some of these molecules is described below. Naturally, speed and efficiency are necessary in the synthesis and purification processes due to the short half-lives of the isotopes involved.

8.2.3.1 ^{11}CO, $^{11}CO_2$, $^{11}CH_3C_5H_{11}$

Carbon-11 is synthesized by irradiating a target of gaseous ^{14}N with 12 MeV protons, whereupon the following nuclear reaction occurs:

$$^{14}N + p \rightarrow {}^{11}C + \alpha$$

Oxygen impurities in the target gas are sufficient to completely oxidize all of the ^{11}C and $^{11}CO_2$ results. The labeled carbon dioxide produced (ca. 10^{-12} moles is produced during a 15 minute irradiation period) can be reduced to ^{11}CO over zinc at 650 K.

^{11}CO or $^{11}CO_2$ can be used as the radiotracer in the PEP experiment or they can be used as precursors for the synthesis of ^{11}C-labeled alkanes.

To produce labeled hexane, the following procedure is adopted [37]. Labeled CO is used in a two-step alkene homologation reaction previously developed in this laboratory [38]. Thus, ^{11}CO is pulsed over a vanadium-promoted Ru/SiO_2 catalyst at 620 K. The temperature is then rapidly reduced to between 360 K and 390 K and 1-pentene is pulsed over the catalyst. Desorptive hydrogenation is then performed. As homologation of shorter alkanes, resulting from cracking of 1-pentene on Ru, also occurs, a range of alkanes is produced from C_1 to C_6 (Fig. 8.4). The hydrocarbons are separated by a process of freezing, flash-heating, and gas chromatography. The desired product, (approximately 1–3 MBq or 10^{-15} moles of $^{11}CH_3C_5H_{11}$) is then frozen out ready for injection. If desired, one of the other labeled alkane fractions can be isolated and injected. The radiolabeled fraction is minute compared with the amount of non-labeled n-hexane that is produced as the total amount of the injected pulse is typically of the order of 10^{-6} moles. The production process (that is the adsorption of ^{11}CO on Ru/SiO_2 and the GC separation) is moni-

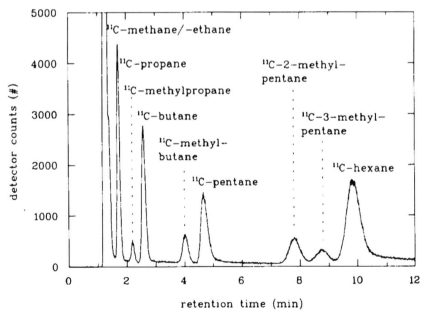

Fig. 8.4 A radio-gas chromatogram of the ^{11}C-labeled products (C_1 to C_6) formed during the C_1/1-pentene homologation reaction used to produce labeled hexanes.

tored by NaI scintillation detectors. The process has been optimized such that pulses of radiolabeled *n*-hexane can be prepared every 45 minutes if desired.

8.2.3.2 ^{13}NO, ^{13}NH$_3$

Nitrogen-13 is produced by irradiating a water target with a beam of 16 MeV protons. During irradiation the following nuclear reaction occurs:

$$^{16}O + p \rightarrow {}^{13}N + \alpha$$

The ^{13}N-labeled species typically exist as nitrates under the conditions of the aqueous target. In order to produce ^{13}N-labeled NH$_3$ or NO, the nitrates formed must be subsequently chemically reduced.

Nitrates are efficiently reduced to nitrogen monoxide in the presence of copper or vanadium [39, 40]. Reduction with copper powder is performed as follows. Copper powder (1 g) is mixed with distilled water (5 mL) and continuously flushed with helium. After addition of the irradiated water (6 mL), 36% HCl solution (0.4 mL) is added to the suspension. The reaction mixture is subsequently brought to the boil. Liberated radioactive gases are passed through a 5 mL NaOH column to remove water and HCl. Gaseous products are trapped on an adsorber at liquid nitrogen temperature. Reduction with VCl$_3$ is performed in a similar way. Irradiated water is added to a solution 5 M in HCl and 0.5 M in VCl$_3$ and the mixture is heated to 373 K. Gaseous products are again dried by passage through a column of NaOH. With both procedures, carrier-free ^{13}NO is produced (i.e. no non-labeled nitrogen species are present).

Production of ^{13}NH$_3$ is performed by reducing the irradiated water (containing ^{13}NO$_3$) over a mixture of DeVarda's alloy and NaOH. The reaction mixture is then flushed with an NH$_3$ (3%)/He flow to transport the liberated ^{13}NH$_3$ and to prevent excessive loss of labeled ammonia on the walls of the gas tubing. Thus, labeled ammonia is not produced carrier-free.

8.2.3.3 ^{15}OO, N$_2$^{15}O

Oxygen-15 can be synthesized by irradiating a target containing gaseous nitrogen with a beam of high-energy deuterons (7 MeV). During irradiation the following nuclear reaction occurs:

$$^{14}N + d \rightarrow {}^{15}O + p$$

About 95% of the labeled products formed exist as mono-labeled oxygen, ^{15}OO. The other 5% exist as N$_2$^{15}O.

During irradiation, products are removed from the target by continuously sweeping it with fresh nitrogen. Radiochemical impurities such as $[^{15}O]O_3$ and $[^{15}O]NO_2$ are removed by passing the effluent over a sodalime trap. After this treatment, only $[^{15}O]O_2$ and $[^{15}O]N_2O$ are present. These products are transported to a zeolite adsorber cooled to 173 K. The N_2O is selectively trapped in this adsorber, and thus a gas flow containing only $[^{15}O]O_2$ remains. In cases where $N_2^{15}O$ is the product of interest, the adsorber is first flushed with helium to remove traces of labeled oxygen and is subsequently flash-heated to release the labeled N_2O [41].

8.3
Applications of PEP in Catalysis

As previously mentioned, the PEP technique enables one to measure concentration profiles of labeled molecules (reactants and/or products) within reactors as a function of axial position and time. Once these profiles have been obtained, they may be compared with the predictions of mathematical models and used for the refinement of physical and chemical properties.

Mathematical models based on mass balance equations under steady-state conditions result in systems of coupled ordinary differential equations (ODE's). These systems are readily numerically soluble. Transient experiments involve systems of coupled partial differential equations (PDE's) as they involve both position and time. These systems are often solved by discretizing one of the variables and then numerically solving the remaining system of ODE's. In PEP experiments in which both the steady-state and transient conditions must be modelled, systems of coupled differential and algebraic equations arise. These systems cannot be solved using normal ODE solvers. For details concerning the solution of these models in combination with parameter estimation the reader is referred to a very lucid paper by Van der Linde et al. [42].

In heterogeneous catalytic reactors, several dynamic phenomena occur simultaneously that may influence the shape of the concentration profiles of reactants and products. These include not only chemical reaction rate parameters (activation energies, pre-exponential factors, sticking probabilities, etc.) but also mass transfer (axial dispersion, convection, and diffusion) and adsorption phenomena.

In this section, we describe how the PEP technique has been applied to studies of all of the above phenomena. This is achieved with the help of several examples.

Firstly, we describe two different techniques that have been used to study mass transport and adsorption properties of alkanes in packed-bed reactors containing zeolites.

8.3.1

Measurement of Mass Transfer and Adsorption Properties of Alkanes in Zeolite Packed-Bed Reactors

Some time ago, we began a project whose ultimate goal was to study the reaction kinetics of the hydroisomerization of *n*-hexane on platinum-containing zeolites using the PEP technique. As with all transient techniques, the contributions of all processes that give rise to changes in concentration profiles are measured simultaneously and thus the individual components must be separated from one another in some fashion.

Hydroisomerization on platinum-containing zeolites proceeds by a bifunctional mechanism at temperatures ranging from 500 to 530 K. In the absence of platinum, no reaction of hexane occurs on acidic zeolites at similar temperatures. Only adsorption/desorption and mass transfer processes occur. Hence, the effects of the adsorption/desorption and mass transfer of alkanes in the bed can be separated from those due to chemical reaction. The parameters obtained from such an investigation, such as the Henry's Law adsorption constant, K_H; the heat of adsorption, ΔH_{ads}; and the diffusion coefficient are not only of fundamental interest in their own right, but are necessary for further investigations and modelling including chemical reactions. This type of experiment is a form of in situ pulse gas chromatography.

In addition to the pulse approach, it seemed worthwhile to extend the PEP technique to include one in which the radiolabel is added by continuous injection into the feed stream of the reactor already operating under steady-state conditions. In this way, tracer exchange experiments can be performed inside the reactor and measured in situ using the PEP detector. This technique, called TEX-PEP, is also described below.

8.3.1.1 The labeled-pulse method (*in situ* tracer pulse chromatography)

Pulse gas chromatography in packed-bed reactors has been used to measure mass transport, adsorption, and reaction kinetic data for the last 40 years. In the classic technique, micromolar quantities of reactant are normally injected into the feed stream of a packed-bed reactor and subsequently the effluent product distribution (a single concentration versus time profile) is analyzed ex situ, after the reactor outlet, using either a thermal conductivity or a flame ionization detector. In some instances, β^--emitting radiotracers (electron-emitters such as carbon-14) are injected and the radiolabeled products are measured via by ionization cell detector again producing a single, ex situ, peak profile. In situ peak detection is not possible since the emitted electrons have insufficient penetrating power to escape from the reactor. See the following references for details concerning this technique [43–45].

Although the classical technique is perfectly acceptable for experiments in which only mass transfer and adsorption, or other fairly predictable phenomena occur, it cannot be extended to more complicated cases in which transient and often highly spatially localized chemical reaction processes also take place. For such situations one must develop an in situ technique.

Here we describe the experimental procedure, the mathematical modelling, and results of such a study in detail [46, 47].

Experimental details

Experiments were performed using a number of different zeolites, such as H-Mordenite, H-Beta, H-ZSM-5, H-ZSM-22, and H-Ferrierite. All zeolites were obtained in their sodium forms and were converted to their acidic forms by overnight ion-exchange using 1 M aqueous NH_4NO_3 (repeated three times) followed by calcination in air at 770 K. Estimates of crystal lengths were obtained from scanning electron micrographs. In order to reduce pressure drop within the reactor, the zeolite crystals were pressed into self-supporting disks and the disks were then broken into smaller pieces and sieved. Sieve fractions of 125–250 µm or 250–500 µm were used.

Experiments were carried out using a typical single-pass flow reactor system as shown schematically in Fig. 8.5. Carrier gases of hydrogen or helium (25–150 mL min^{-1}) were supplied by mass flow controllers. The reactor itself consisted of a quartz tube (4 mm ID) and contained ca. 300 mg of pelletized acidic zeolite. The reactor bed was 42 mm in length. Pressure gauges were placed immediately before

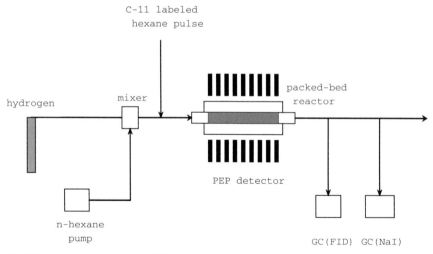

Fig. 8.5 A schematic diagram of the single-pass flow reactor set-up and the PEP detector. Ex situ analysis of (un)labeled reactants and products is also performed by on-line GC and radio-GC.

and after the reactor and the measured pressure drops were used to measure the bed voidage, ϵ_b, according to the Ergun equation. The zeolites were dried overnight in flowing hydrogen (150 mL min^{-1}) at 670 K prior to the experiment.

The reactor and oven were placed in between the PEP detector banks. A pulse containing ca. 10^{-15} moles of labeled n-hexane ($^{11}CH_3C_5H_{11}$) was injected into the reactor feed stream and the concentration profiles were measured using the PEP detector.

Fig. 8.6(a) shows a typical PEP plot obtained when a pulse of $^{11}CH_3C_5H_{11}$ was injected into a stream of hydrogen (150 mL min^{-1}) at 423 K flowing through a bed of H-Mordenite (250–500 μm). The spectrum is a two-dimensional representation of the measured activity-time-position profile. The residence time is plotted against the axial displacement within the bed. The activity measured at that position and time is represented by the grey scale according to a linear colour bar (also shown). Fig. 8.6(b) shows horizontal cross-sections that were taken from the data set shown in Fig. 8.6(a). These cross-sections represent the measured activity as a function of axial position within the reactor at different retention times.

Data analysis (modelling)

As mentioned above, these experiments represent an in situ version of the tracer pulse gas chromatographic method. Since the physicochemical mass transfer processes in packed beds are similar in these experiments, the same data analysis methods can be applied. The difference here is that the experimental data generated include 17 different concentration-time profiles measured at regular (3 mm) intervals through the bed, rather than a single profile measured after it.

A packed-bed reactor containing particles of zeolite crystals represents a biporous adsorbent system. The use of such a system in gas-solid chromatography has been extensively discussed and its behavior has been modelled; see, for example [48–50]. Several methods of data analysis have been proposed, including analysis in the time domain, the Fourier domain, and the Laplace domain. Time domain analysis has been proposed as giving the most reliable results [50, 51].

Time domain models are based on the time-dependent continuity equations for the bed, macropores, and micropores. For some cases, such as that in which an injected delta pulse is employed, these systems can be solved analytically [50]. Initially, the PEP data were analyzed by numerical fitting to one such analytical solution [46]. Further information was then obtained by applying the method of moments to the fitted profiles. However, analysis based on numerical solution of the time-dependent mass balance equations is necessary when one wishes to proceed to more complex situations, for example, to include reaction kinetics. Thus, the modelling necessary is described here in some detail.

The mathematical model used to calculate the concentration profiles of (labeled) molecules in a biporous adsorber was originally presented by Haynes [52] to describe the transport through a chromatographic column. The model describes

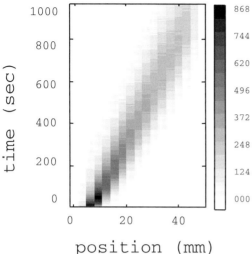

Fig. 8.6(a) A typical PEP plot obtained following injection of a $^{11}CH_3C_5H_{11}$ pulse into a stream of hydrogen (150 mL min^{-1}) at 423 K flowing through a bed of H-Mordenite (250–500 µm). The residence time (y-axis) is plotted against the axial displacement (x-axis) within the bed. The activity measured at that position and time is represented by the grey scale according to a color bar (also shown).

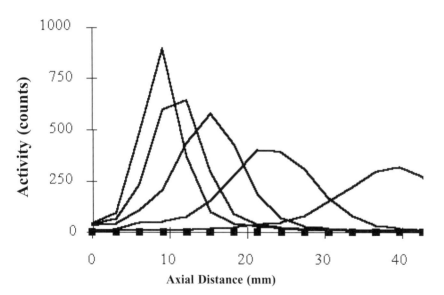

Fig. 8.6(b) Horizontal cross-sections taken from the data set shown in Fig. 8.6(a). These cross-sections represent the measured activity as a function of axial position in the reactor at different times.

mass transport in four phases: the gas flowing through the bed between the macroparticles usually denoted as the "fluid phase", a stagnant film surrounding the macroparticles, the macropore gas phase, and the micropore gas phase. Conventionally, this model describes alkane transport in the micropores (of molecular dimensions) using a rather artificial gas-phase concentration and an immobile adsorbed phase. It should be noted that most experimental studies report the effective diffusion coefficient in the micropores in the form of an adsorbed phase diffusion constant. In the model, each phase is represented by its own coordinate system: x represents the radial coordinate of the microparticles, y the radial coordinate of the macroparticles, and z the linear coordinate along the packed bed. The gas-phase concentrations of hexane in hydrogen are position- and time-dependent and are represented by C with a subscript that indicates the phase. Transport of molecules between the different phases is achieved through the boundary conditions of the mass transport equations of the individual phases.

Mass transport in the spherically-symmetrical microparticle is described by diffusion in radial coordinates:

$$\epsilon_x(1 + K_a)\frac{\partial C_x}{\partial t} = D_x \left(\frac{\partial^2 C_x}{\partial x^2} + \frac{2}{x} \cdot \frac{\partial C_x}{\partial x} \right) \tag{1}$$

The microparticle porosity is represented by ϵ_x. The adsorption equilibrium constant K_a is defined as the number of adsorbed molecules in a microparticle divided by the number of gas-phase molecules in the microparticle. It is assumed that adsorption is instantaneous and reversible and that the adsorption isotherm is linear. The temperature dependence of the adsorption equilibrium is controlled by the adsorption energy ΔH according to:

$$K_a = K_{a,\infty} e^{-\Delta H/RT} \tag{2}$$

where $K_{a,\infty}$ is the adsorption equilibrium constant at infinite temperature, R is the gas constant, and T is the absolute temperature.

D_x is the microparticle gas-phase effective diffusion coefficient with an activation energy E_{act}.

$$D_x = D_{x,\infty} e^{-E_{act}/RT} \tag{3}$$

$D_{x,\infty}$ is the gas-phase micropore diffusion coefficient at infinite temperature. The relationship between D_x and the effective adsorbed phase diffusion coefficient D_{micro}, often referred to in the literature, is:

$$D_x = \epsilon_x(1 + K_a) D_{micro} \tag{4}$$

In practical situations, $K_a \gg 1$. As a result, D_x has an activation energy that is equal to the sum of the heat of adsorption and the activation energy for gas-phase diffusion. The boundary conditions (BC^s) of Eqn. (1) assume a spherically symmetrical concentration profile in the microparticle, which gives:

$$\frac{\partial C_x}{\partial x}(0,t) = 0 \tag{5}$$

Empty microparticles at $t = 0$ are expressed as:

$$C_x(0,t) = 0 \tag{6}$$

Coupling between the gas-phase concentration and the macropores is provided by:

$$C_x(R_x,t) = C_y(y,t) \tag{7}$$

The gas-phase concentration at the outer microparticle boundary is equal to the macroparticle gas-phase concentration at its position in a macroparticle.

Mass transport in the spherically symmetrical macroparticles is described by diffusion in the macropores:

$$\varepsilon_y \frac{\partial C_y}{\partial t} = D_y \left\{ \left(\frac{\partial^2 C_y}{\partial y^2} + \frac{2}{y} \frac{\partial C_y}{yt} \right) + \frac{3(1 - \varepsilon_y)}{D_y R_x} \cdot N_{R_y} \right\} \tag{8}$$

This equation differs from the transport equation for the microparticles (1) in that it includes an additional source term on the right-hand-side that accounts for exchange between the macropores and micropores. R_x represents the radius of the microparticle, ε_y represents the macroparticle porosity, and D_y is the effective macroparticle diffusion coefficient, which is defined as the molecular diffusion coefficient of hexane in hydrogen divided by the macroparticle tortuosity:

$$D_y = \varepsilon_y \frac{D_m}{\tau_y} \tag{9}$$

In principle, the macropore diffusion coefficient is made up of two contributions: molecular and Knudsen diffusion. The latter is negligible since the mean free path of molecules was found to be significantly smaller than the macropore diameter.

The temperature dependence of D_m [53] is calculated as:

$$D_m = 0.3 \cdot 10^{-4} \left(\frac{T}{T_{room}} \right)^{1.8} \tag{10}$$

where $T_{room} = 293$ K.

The mass transport between the microparticle gas phase and the macroparticle gas phase is achieved with the N_{R_x} term. The rate of mass transport is proportional to the gas-phase concentration gradient at the boundary:

$$N_{R_x} = -D_x \frac{\partial C_x}{\partial x} \bigg|_{R_x} \tag{11}$$

The boundary conditions are (empty macroparticle at $t = 0$; spherically symmetrical concentration distribution):

$$C_y(y,0) = 0 \tag{12}$$

$$\frac{\partial C_y}{\partial y}(0,t) = 0 \tag{13}$$

Finally, mass transfer through the fluid phase in the column is described by:

$$\varepsilon_z \frac{\partial C_z}{\partial t} = D_z \frac{\partial^2 C}{\partial z^2} - v_{sup} \frac{\partial C_z}{\partial z} + \frac{3(1 - \varepsilon_z)}{R_y} \cdot N_{R_y} \tag{14}$$

where R_y is the radius of the macropore. The porosity of the packed bed is ε_z. Diffusion in the fluid phase has two contributions:

$$D_z = \varepsilon_z \left(0.5 \frac{D_m}{\tau_z} + 0.5 \cdot 2 R_y v_{int} \right) \tag{15}$$

The first term of (15) describes molecular diffusion, with τ_z being the tortuosity of the bed, while the second term describes mixing in vortices by eddy diffusion.

The superficial velocity of the carrier gas v_{sup} in an empty tube follows from the flow rate of the carrier gas and the cross-sectional area of the column. The interstitial velocity v_{int} is defined as v_{sup}/ε_z.

The mass transfer between the macroparticles and the fluid phase, described by the last term in (16), involves transport through a stagnant layer/film surrounding the macroparticles. The concentration in the macropores at the macroparticle boundary therefore equals the concentration in the packed bed phase after correction for the concentration gradient over the film surrounding the macroparticle:

$$C_y(R_{y,t}) = C_z(z,t) + N_{R_y}/k_f \tag{16}$$

Here, k_f is the film-transfer constant, which is calculated from:

$$k_f = \frac{D_m Sh}{2 R_y} \tag{17}$$

where Sh is the Sherwood number of the macroparticles. For Reynolds numbers larger than 3, the Sherwood number can be calculated from $Sh = 2 + 1.1 Re^{0.6} Sc^{0.33}$ [50], with Sc being the Schmidt number of hydrogen. For Reynolds numbers smaller than 3, Sh is equal to 2 and thus k_f is given by:

$$k_f = \frac{D_m}{R_y} \tag{18}$$

Consequently, the rate of mass transport between the fluid phase and the outer macroparticle boundary is given by:

$$N_{R_y} = -D_y \left. \frac{\partial C_y}{\partial y} \right|_{R_y} \tag{19}$$

The first boundary condition for mass transfer through the column (20) is the initial condition of an empty column before injection, i.e. at $t = 0$:

$$C_z(z,0) = 0 \tag{20}$$

Since the modelled column has a finite length, a mass balance at both ends of the column is used to derive additional boundary conditions:

$$v_{sup} C_{z,0^-} - D_{z,0^-} \frac{\partial C_{z,0^-}}{\partial z} = v_{sup} C_{z,0^+} - D_{z,0^+} \frac{\partial C_{z,0^+}}{\partial z} \tag{21}$$

The index 0^- indicates directly before, 0^+ directly after the column entrance.

Since the diffusivity in the packed bed is much larger than in the empty part of the tube preceding it (no mixing contribution), the term with $D_{z,0^-}$ is negligible and equation (21) can be reformulated as the Neumann boundary condition:

$$\frac{\partial C_{z,0^+}}{\partial z} = \frac{v_{sup}}{D_{z,0^+}} (C_{z,0^+} - C_{z0^-}) \tag{22}$$

A similar argument applied to the column exit gives:

$$C_{z,1^-} = C_{z,1^+} + \frac{D_{z,1^-}}{v_{sup}} \frac{\partial C_{z,1^-}}{\partial z} \tag{23}$$

The index 1^- indicates directly before, 1^+ directly after the column exit. However, application of (23) is not possible since $C_{z,1^+}$ is not modelled.

In their study of chromatographic columns filled with 5 Å and 10x molecular sieve pellets, Raghavan and Ruthven assumed Danckwerts boundary conditions [54] instead of (22) and (23). However, Danckwerts BC[s] are only justified for steady-state conditions [55, 56]. Application of Danckwerts boundary conditions results in the use of equation (22) at the column entrance, but it assumes no concentration jump at the column exit.

$$\frac{\partial C_{z,1^-}}{\partial z} = 0 \tag{24}$$

Numerical evaluation of the model

The model described in the previous section is a set of partial differential equations (PDE's) coupled through their boundary conditions combined with algebraic equations to form a Differential Algebraic Equation (DAE). Such systems can be solved numerically with the method of lines [57] yielding concentrations in the different phases as functions of position and time.

In gas chromatography, the fluid phase solution is normally calculated at the end of the column following a pulse injection at the entrance.

In contrast, the PEP detector images all phases within the column: the fluid phase, the macroparticle phase, and the microparticle phase all contribute to the measured PEP profiles. Therefore, all phase solutions have to be converted into one PEP profile. This conversion is achieved by volume averaging. The contributions of each different phase are summed to give the total measured concentration at each grid point along the bed. The average concentration in one microparticle is given by:

$$\langle C_x \rangle (y,t) = \frac{3}{R_x^3} \int_0^{R_x} (K_a + 1) C_x x^2 dx \tag{29}$$

Similarly, the resulting average macroparticle is:

$$\langle C_y \rangle (z,t) = \frac{3}{R_y^3} \int_0^{R_y} \left[\varepsilon_y C_y (y,t) + (1 - \varepsilon_y) \langle C_x \rangle \right] y^2 dy \tag{30}$$

The resulting average concentration at a point z in the reactor becomes:

$$\langle C_z \rangle (z,t) = \left[\varepsilon_z C_z (y,t) + (1 - \varepsilon_z) \langle C_y \rangle \right] \tag{31}$$

Although PEP measures the average concentration in a volume ΔV associated with the position resolution of the detection system, the concentration in the middle of this volume element is used for the average concentration therein. The integrals needed for the volume averaging have to be solved numerically because of the discrete character of the solution.

Results

Only sample solutions are shown here. Figure 8.7 shows the measured and the fitted chromatograms of labeled hexane in H-Mordenite using optimized values for K_H, ΔH, and the diffusivities at 443 K. The relative contributions of different mechanisms to the peak broadening have been calculated and are listed in Tab. 8.1. Apparently, axial dispersion in the bed, mass transfer through the stagnant film, and mass transfer in the macro- and micropores all contribute to the observed peakwidth for hexane in H-Mordenite. Similar results were found for HZSM-5.

Figure 8.8 shows the measured profiles along with the modelled profiles using optimized values for K_H, ΔH, and the diffusivities on H-ZSM-22 at 413 K. As Tab. 8.1 clearly shows, micropore diffusion is the dominant factor contributing to peak

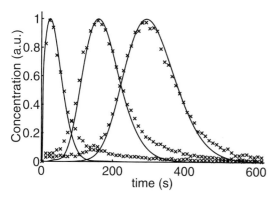

Fig. 8.7 Normalized measured PEP profiles (x) following an $n\text{-}^{11}CH_3C_5H_{11}$ pulse in hydrogen flowing through H-Mordenite at 443 K, $v_{sup} = 0.30$ m s^{-1}. Fitted profiles (−) with optimized values for K_H and ΔH and diffusivities are also shown.

Tab. 8.1 The relative contributions (%) to the second moment σ^2 at the end of the column ($z = 1$).

Zeolite	v_{sup} (m s^{-1})	T (K)	D_L (%)	Film (%)	Macropores (%)	Micropores (%)
H-Mordenite	0.30	443	34	13	22	31
H-ZSM-5	0.31	463	46	18	34	0.5
H-ZSM-22	0.28	423	19	9	20	52
H-Ferrierite	0.27	403	22	10	23	45

broadening. Thus, an entirely different behavior is observed in this small-pore zeolite compared with the medium-pore zeolites. The small-pore zeolite H-Ferrierite behaves similarly to H-ZSM-22.

Important conclusions were drawn from the combined results of experiments and model calculations concerning the potential of positron emission profiling for studying reaction kinetics of n-hexane under steady-state conditions in biporous packed beds. For the medium-pore zeolites, experimental conditions cause axial dispersion to dominate the broadening of the concentration profiles. However, axial dispersion is only slightly influenced by molecular properties. The gas-phase diffusion coefficients of n-hexane and its isomers are similar.

This stresses the necessity that the reactant and the reaction product(s) must have different mass transport properties if they are to be "resolved" in PEP experiments. In a biporous bed, this is probably only achievable if there is a significant difference in the effective microcrystalline diffusion coefficients of the reactant and reaction product(s). Even when this condition is met, the shape of the PEP profiles is only directly influenced by the reaction when micropore diffusion is the dominant contribution in the second moment (the standard deviation) of the PEP profiles. This criterion is similar to that necessary for successful measurement of micropore diffusion constants by chromatographic means previously noted by Post [58]. The criterion states that the micropore diffusion time constant,

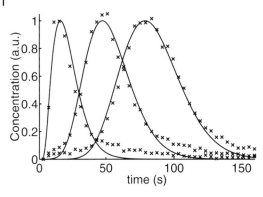

Fig. 8.8 Normalized, measured PEP profiles (x) following an n-^{11}CH$_3$C$_5$H$_{11}$ pulse in hydrogen flowing through H-ZSM-22 at 423 K, $v_{sup} = 0.28$ m s^{-1}. Fitted profiles (–) after optimization of K_H, ΔH, and the micropore diffusion coefficient.

R_x^2/D_{micro}, should be greater than 100 s. According to measurements made here, this criterion would be met for medium-pore zeolites such as H-Mordenite or H-ZSM-5 under realistic operational conditions if the zeolite crystals were to have diameters greater than 100 μm. In small-pore zeolites, such as H-ZSM-22 and H-Ferrierite, micropore diffusion determines the shape of the PEP profiles to a large extent even when small pelletized crystals are used. Thus, with such zeolites, measurement of reaction kinetics under steady-state conditions may well be possible using the in situ PEP technique.

8.3.1.2 By leak injection: tracer exchange positron emission profiling (TEX-PEP)

The extension of the PEP technique to include continuous injection of the radiolabel is equivalent to developing an in situ form of the SSITKA technique. In SSITKA, feed molecules are exchanged by isotopically-labeled ones, for example ^{12}C/^{13}C or ^{16}O/^{18}O, and the kinetics of the exchange process are followed ex situ, typically by means of mass spectrometry. For more information on this technique the reader is referred to a review by Shannon and Goodwin [59].

The principle of **T**racer **EX**change-**PEP** (TEX-PEP) is outlined below. Again, we describe the method using measurements of the adsorption and diffusion properties of alkanes in zeolites as an example.

Experimental details

The experimental set-up is essentially the same as that discussed above in Section 8.3.1.1. The only modifications required are those necessary to change from a pulsed to a "constant leak" injection of the radiolabeled molecules. This is achieved using a syringe pump and a gas-tight syringe that permits feeding of the collected gas at a pumping speed between 0.001 and 20 mL min^{-1}. At the start of the experiment, the contents of the syringe (in this case 10^{-15} moles of ^{11}C-labeled 2-methylpentane plus submicromolar quantities of unlabeled 2-methylpentane) are continuously injected at a low flow rate (F_2) into the steady-state stream (F_1). Small leak flow rates (of the order of 0.1–0.5 mL min^{-1}) are used in order to minimize the influence of switching and to optimize the radioactive signal obtained from the catalyst bed ($F_2/F_1 \leq 0.05$) [60].

As dynamic processes occur, the unlabeled fraction inside the zeolite bed exchanges with the labeled fraction. This process is monitored as a function of time at each detector position along the reactor axis (every 3 mm). When the radioactive signal from the catalyst particles reaches a constant value, the exchange is assumed to be complete and the injection of the radiolabel is stopped. The subsequent re-exchange process is also monitored.

Figure 8.9 shows an experimental data set measured during the exchange of labeled 2-methylpentane at 413 K in a bed of large silicalite crystals previously equilibrated in a feed of 2-methylpentane in hydrogen. Although curves were measured at 17 different positions, only ten are shown in the figure for clarity. It can be seen that after a certain time, a sequential increase in the count rate is observed in all detector positions. The monitored intensity is somewhat lower at the first and the last detector positions than at the other positions as the zeolite column does not com-

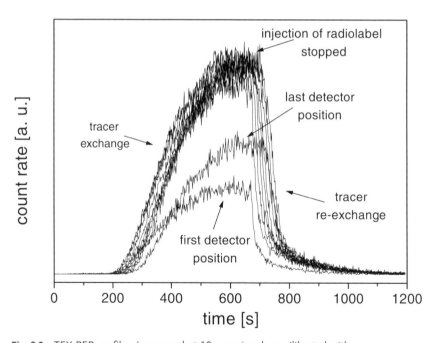

Fig. 8.9 TEX-PEP profiles (measured at 10 different positions along the reactor axis) during the exchange of [^{11}C]2-methylpentane in silicalite crystals previously equilibrated with [^{12}C]2-methylpentane at 373 K. Re-exchange profiles (after label feed stopped) are also shown.

pletely fill these two outer positions. Figure 8.9 also shows the re-exchange of the labeled molecules that occurs after the injection of the radiolabel is stopped. Generally, the re-exchange process yields more reliable information because the termination of the injection occurs more abruptly than initiation and therefore its position can be more accurately determined. The switching at the entrance of the catalyst bed occurs on time scales shorter than the experimental time resolution (ca. 2 s). Temporal resolution is determined by the integration time that is necessary to produce statistically reliable count rates.

By fitting appropriate models to the time evolution of the tracer exchange at the various positions along the reactor bed, information can be obtained on the rates of the elementary process(es) that is/are rate-determining. As discussed above, this might be a transport process such as macropore diffusion in the catalyst pellet or micropore diffusion in the pores of a zeolite, or an adsorption/desorption process. It should be emphasized that the elementary process that is rate-determining for the tracer exchange is not necessarily identical to that which determines the rate of the catalytic reaction under steady-state conditions.

Modelling

The model described in detail in Section 8.3.1.1 can also be used to describe TEX-PEP experiments with only a few modifications [60]. As large crystals are used instead of pelletized smaller crystallites, the packed-bed is monoporous rather than biporous; no macropores are present.

The mass balance equation given previously (Eqn. 1), neglecting film diffusion, becomes:

$$\frac{\partial C_z}{\partial t} = D_{ax}\frac{\partial^2 C_z}{\partial z^2} - v_{int}\frac{\partial C_z}{\partial z} + \frac{3(1-\varepsilon)}{\varepsilon R_c}N_{Rc} \tag{32}$$

As before, mass transfer across the crystallite boundary is assumed to be Fickian and therefore the intracrystalline flux can again be described as in Eqn. 11.

$$N_{Rc} = -D_c\frac{\partial C_x}{\partial x}\bigg|_{r=Rc} \tag{33}$$

The boundary conditions are derived from a mass balance at both ends of the column, similar to Eqn. 22.

$$\frac{\partial C_{z,0^+}}{\partial z} = \frac{v_{int}}{D_{ax}}(C_{z,0^+} - C_{z0^-}) \tag{34}$$

The concentration just before the column entrance is defined in a TEX-PEP experiment by the Heaviside step function.

$$C_{z,0^-}(t) = C_0, \quad t \le 0$$

$$C_{z,0^-}(t) = 0, \quad t > 0$$

(35)

Mass transfer inside the zeolite crystallites is assumed to follow Fick's second law, hence the mass balance equation inside the crystallites (assumed to be spherical) becomes:

$$\frac{\partial C_x}{\partial t} = D_c \left[\frac{\partial^2 C_x}{\partial t^2} + \frac{2}{x} \frac{\partial C_x}{\partial x} \right]$$

(36)

For symmetry reasons, an absence of mass transfer across the center of the sphere has been assumed.

$$\left. \frac{\partial C_x}{\partial x} \right|_{x=0} = 0$$

(37)

Equilibrium is assumed between the gas phase and adsorbed phase at the outer boundary:

$$C_x(x = R_c, z, t) = K_{ads} \times C_z(z,t)$$

(38)

At time $t = 0$, the bed and crystals are saturated with adsorbate and the entire system is at equilibrium. Hence, initially, no concentration gradients exist over the crystallites. Every point in the crystal is in equilibrium with the gas phase, thus:

$$C_z(z, t = 0) = C_0$$

(39)

$$C_x(x, z, t = 0) = K_{ads} \cdot C_0$$

(40)

As in the previous example, the system of equations must be discretized and solved numerically, using, for instance, the method of lines. The solution produces equations for C_z and C_x as a function of time and bed/crystal grid point. Again, volume averaging is necessary. This is achieved using Eqns. 41 and 42.

$$\langle C_x(z,t) \rangle = \frac{3}{R_x^3} \int_0^{R_c} C_x(x,z,t) * x^2 dx$$

(41)

$$\langle C_{tot}(z,t) \rangle = \varepsilon C_z(z,t) + (1 + \varepsilon) * \langle C_x(z,t) \rangle$$

(42)

Results

The diffusivity of 2-methylpentane in silicalite was chosen as a test case with which to evaluate the TEX-PEP method as published literature data for this system are in relatively good agreement. These data, taken from [61–63], are presented in

Tab. 8.2 Comparison of experimental data for the diffusivity of 2-methylpentane in MFI-type zeolites.

Ref. no.	Method	Temp. [K]	D [cm² s⁻¹]	D (T = 413 K) [cm² s⁻¹]	E_A [kJ mol⁻¹]
[61]	gravimetry	297–338	5×10^{-10} T = 338 K	5×10^{-9}	35
[62]	gravimetry	373–473	2×10^{-8} T = 423 K	1.5×10^{-8}	46
[63]	transient analysis of products (TAP)	475–598	3×10^{-8} T = 475 K	1.2×10^{-8}	24
this work	TEX-PEP	373–523	1.0×10^{-8} T = 413 K	1.0×10^{-8}	29 ± 3

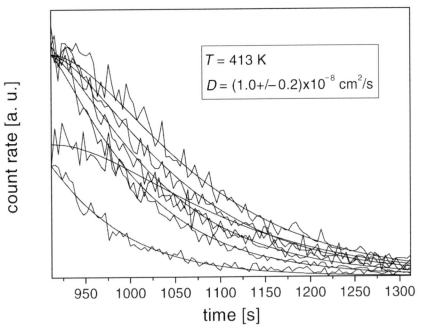

Fig. 8.10 TEX-PEP re-exchange profiles for 2-methylpentane in silicalite, measured at six positions along the reactor bed (at 6 mm intervals), are shown together with fitted profiles and the calculated micropore diffusivity.

Tab. 8.2, along with information on the experimental technique and the temperature range used. The data have been extrapolated to 413 K to enable direct comparison. The reported diffusivities deviate by a factor of 2–3 in the temperature range of interest. This is in contrast to deviations of several orders of magnitude commonly encountered for many other systems.

Figure 8.10 shows an example of experimental tracer re-exchange curves determined at 413 K. Experimental data for six different positions equally distributed along the reactor bed are shown together with the simulated curves derived using the above model. As mentioned before, the concentration curves for the first and the last position of the zeolite bed start at somewhat lower values, as the column did not completely fill these detector positions. The calculated diffusivity value is included in Tab. 8.2. The temperature dependence was also measured and an activation energy for diffusion of 29 ± 3 kJ mol^{-1} was calculated.

The values for the diffusivity and the activation energies are compared with literature data in Tab. 8.2. The diffusivity data measured in all four investigations are in excellent agreement with each other. The measured activation energy was also consistent with those reported previously [61, 63].

8.3.2
Measurement of the Reaction Kinetics of CO Oxidation on Pt/Ceria/Alumina Using ^{11}CO

The above examples show how the PEP radiolabeled tracer technique enables in situ measurement of transport and adsorption properties. In this example, experiments were performed under conditions where reaction rates were rate-limiting. In this instance, mass transport (other than convection) played no role and mathematical models based on elementary steps such as adsorption, desorption, and surface reaction could be used to simulate these profiles. Numerical fitting of experimental data to models produced refined kinetic parameters [64].

Experiments were performed using simulated three-way exhaust catalysts, ceria-promoted Pt supported on γ-alumina. Small particles and high space velocities were used so that transport phenomena (other than convection) were excluded. These conditions are typically observed during the so-called "cold start" period of an automobile.

8.3.2.1 Experimental details

Standard catalysts used consisted of highly dispersed platinum on γ-alumina containing 0.6 wt.% highly dispersed ceria. Table 8.3 shows a list of their properties.

Tab. 8.3 Properties of the Pt/CeO$_2$/Al$_2$O$_3$ catalyst

Support	Alumina/Ceria
Particle size	30–80 mesh
BET surface area	111 m^2 g^{-1}
Pore volume	0.56 mL g^{-1}
Ceria loading	0.6 wt.%
Ceria surface	Highly dispersed on alumina surface
Platinum loading	0.12 wt.%
Platinum dispersion	84%

Experiments were carried out within a single-pass, packed-bed microreactor constructed from a 3/8″ (9.5 mm) stainless steel tube enclosed within a cylindrical, 15 mm diameter silver heating block. The catalyst bed (3.9 g catalyst) was 13.5 cm long with a diameter of 0.70 cm.

Steady-state experiments were performed by continuously passing CO/O$_2$/Ar mixtures [1.0% (*v/v*) CO, 0.5% (*v/v*) O$_2$ in Ar] at 40 mL min^{-1} at 413 K. Pulses of the same gas mixture also containing ca. 50 MBq (0.15 pmole) of ^{11}CO or ^{11}CO$_2$ were subsequently injected. Products were analyzed ex situ using a gas chromatograph equipped with both thermal conductivity and NaI scintillation detectors, and by a quadrupole mass spectrometer.

8.3.2.2 Modelling

The oxidation of CO on Pt has been extensively studied both on single crystals and on supported catalysts. From these studies, the following is known about this reaction [65–71]:

The reaction is:

- first order in molecular oxygen;
- negative first order in CO;
- at low temperatures, the surface is largely covered by CO, thus the rate of oxygen adsorption is limiting;
- oxygen adsorption is very fast and dissociative;
- surface reaction of CO and O to produce CO$_2$ is rapid, leading to immediate desorption of CO$_2$ and regeneration of two adsorption sites.

This information was used to construct the following reaction mechanism:

1) $CO + Pt \leftrightarrow PtCO$
2) $O_2 + 2\,Pt \rightarrow 2\,PtO$
3) $PtCO + PtO \rightarrow 2\,Pt + CO_2$
4) $CO_2 + A \leftrightarrow ACO_2$

where A is an adsorption site on alumina necessary to account for hold-up due to adsorption on the carrier. Note that the formation of carbon dioxide is assumed to be irreversible.

The concentrations of the reactants, intermediates, and products, in either the gas or the adsorbed phase, are functions of both axial position and time. The mass balance equations for each of these components for $t > 0$ and for $0 < x < L$ cm are:

In the gas phase:

$$\frac{\partial(CO)}{\partial t} + v * \frac{\partial(CO)}{\partial x} = -\frac{(1-\varepsilon)}{\varepsilon} * R_1 \tag{43}$$

$$\frac{\partial(O_2)}{\partial t} + v * \frac{\partial(O_2)}{\partial x} = -\frac{(1-\varepsilon)}{\varepsilon} * R_2 \tag{44}$$

$$\frac{\partial(CO_2)}{\partial t} + v * \frac{\partial(CO_2)}{\partial x} = \frac{(1-\varepsilon)}{\varepsilon} * (R_3 - R_4) \tag{45}$$

In the adsorbed phase:

$$\frac{\partial(PtCO)}{\partial t} = (R_1 - R_3) \tag{46}$$

$$\frac{\partial(PtO)}{\partial t} = 2 * R_2 - R_3 \tag{47}$$

$$\frac{\partial(ACO_2)}{\partial t} = R_4 \tag{48}$$

$$(PtCO) + (PtO) + (Pt) = Pt_{max} \tag{49}$$

$$(ACO_2) + (A) = A_{max} \tag{50}$$

where R_x is the rate equation for step $\{x\}$, for example:

$$R_1 = k_{1,f}[CO][Pt] - k_{1,b}[PtCO] \tag{51}$$

A pulse containing a negligibly small amount (2×10^{-12} moles) of labeled ^{11}CO or $^{11}CO_2$ does not alter the total concentration of the steady-state; only exchange between components with and without ^{11}C-labeled carbon atoms takes place. The above system of equations can first be solved for the steady-state condition by omitting all terms involving derivatives in time, as these quantities are equal to zero in the steady-state. The resulting system of coupled ordinary and algebraic equations can then be solved yielding the concentration profiles of each component as a function of axial position within the reactor under steady-state conditions.

To solve the time-dependent problem, in which a transient Gaussian-shaped pulse is superimposed on the steady-state solution, the above system of equations,

Reaction step no.	Parameter	
1	S_{co}	CO sticking probability
−1	E_a/R	activation energy for CO desorption
−1	K_0	pre-exponential factor for CO desorption
2	S_{O2}	O_2 sticking probability
3	E_a/R	reaction activation energy
3	K_0	reaction pre-exponential factor

Tab. 8.4a Necessary kinetic reaction parameters.

including time derivatives, must be solved. In addition, mass balance equations must be added for all ^{11}C-labeled components. The resultant system of hyperbolic partial differential equations can be solved numerically by the method of lines [57].

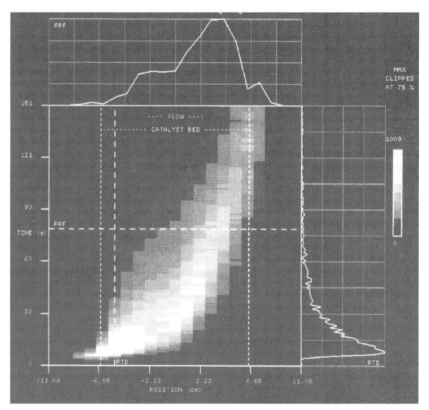

Fig. 8.11(a) PEP image of a [11]CO pulse experiment on Pt/ceria/alumina under steady-state conditions at 413 K. Profiles are also shown at fixed time (80 s) (top) and at fixed position (−5 cm) (right-hand-side).

Initial estimates for the various kinetic parameters are necessary, and these were taken from the previous work of Lynch et al. (see Tab. 8.4a) [72].

8.3.2.3 Results

Figure 8.11(a) shows the PEP image of a ^{11}CO pulse experiment on Pt/ceria/alumina under steady-state conditions at 413 K. Profiles are also shown at fixed time (80 s) (top) and at fixed position (–5 cm) (right-hand-side).

Figure 8.11(b) shows cross-sections of Fig. 8.11(a) taken at constant positions.

As shown in Figs. 8.11(a) and 8.11(b), the profiles obtained under conditions where rate kinetics are rate-limiting are similar in shape and behavior to those shown above for cases in which mass transport and adsorption were limiting. Thus, great care is necessary when designing, performing, modelling, and interpreting the findings from these experiments.

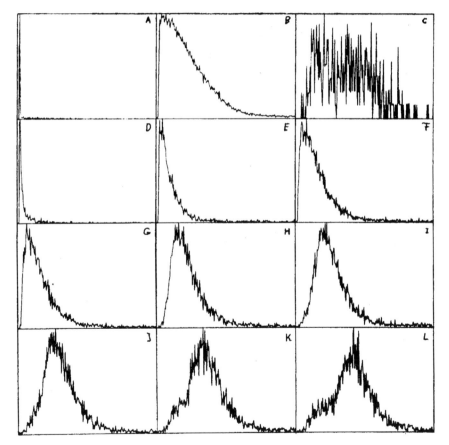

Fig. 8.11(b) Horizontal cross-sections (constant position) of Fig. 8.11(a) are shown at several different retention times.

Tab. 8.4(b) Comparison between determined parameters and those in previous literature [72].

Reaction step no.		Determined value (this study)	Literature value [72]
1	S_{CO}	8×10^{-5}	2×10^{-5}
−1	E_a/R	9×10^3 K	9×10^3 K
−1	k_0	1.8×10^{10} mol m^{-3} s^{-1}	4×10^9 mol m^{-3} s^{-1}
2	S_{O2}	3.6×10^{-7}	9×10^{-9}
3	E_a/R	8×10^3 K	8×10^3 K
3	k_0	5.8×10^{13} s^{-1} m^{-1}	5.8×10^{13} s^{-1} m^{-1}

Fitting the profiles to the model described above yielded refined kinetic parameters under reaction conditions (see Tab. 8.4b). As shown in the table, excellent agreement was achieved with previous measurements made at reduced pressure [72]. Such agreement shows that the so-called "pressure gap" is not always prohibitive and that information gained under UHV conditions can be relevant to "real reaction conditions". In any case, such data are necessary to provide essential initial estimates that can be refined by in situ techniques such as PEP.

8.4
Other (Potential) Applications of PEP in Catalysis Research

This chapter has shown that PEP can be used to acquire concentration profiles as a function of axial position and time within reactors under normal operating conditions. These data can be fitted using mathematical models to obtain parameters for mass transport, diffusion, adsorption, and reaction kinetics. Such data are vital for the understanding, prediction, and design of chemical reactions and reactors.

The examples discussed here are in no way exhaustive. Many other reactions and reaction phenomena have already been investigated using PEP.

The ability to measure concentrations in situ with spatial resolution has proven very useful in studying localized phenomena such as deactivation through the deposition of carbonaceous species on metals [73], and spillover behavior of oxygen from Pt to ceria [64] and onto alumina during $N_2^{15}O$ decomposition on Pt/alumina [74].

Now that the diffusion and adsorption behavior of alkanes on Pt/H-zeolites is understood, we are now modelling reaction kinetics of the hydroisomerization reaction (and related steps) under real conditions [75].

The TEX-PEP measurements described above gave a measure of the diffusivity and adsorption behavior of alkanes in a zeolite-containing reactor under steady-state process conditions. This technique is now being extended to measure such parameters in binary and other mixtures making it one of the few techniques capable of measuring such data. Efforts to measure reaction kinetics simultaneously are also underway.

Previous studies of the selective, low-temperature oxidation of ammonia to nitrogen and water on noble metals (Pt, Ir) showed that these catalysts are susceptible to rapid deactivation due to irreversible adsorption of reaction "intermediates" such as NH_x species [76]. PEP experiments are now underway using either $^{13}NH_3$ or ^{15}OO to further understand this behavior.

Finally, the so-called Matros "reverse-flow reactor" has been shown to offer many beneficial features related to heat transport, enhanced reactant conversions, and directed product selectivities [77]. PEP will be used to monitor concentration profiles within reverse flow reactors in order to supply more kinetic information for use in experimental design and predication.

List of Symbols Used

ε_x	microparticle porosity, m³ micropores/m³ microparticle ( )
ε_y	macroparticle porosity, m³ macropores/m³ macroparticle ( )
ε_z	packed bed porosity, (1 - m³) macroparticles/m³ column ( )
$\langle C_{x,y,z} \rangle$	average concentration in microparticle, macroparticle, or fluid phase (moles m⁻³)
C_x	microparticle gas-phase concentration (moles m⁻³)
C_y	macroparticle gas-phase concentration (moles m⁻³)
C_z	fluid-phase gas concentration (moles m⁻³)
$C_{z,0^-}$	fluid-phase concentration before column entrance (moles m⁻³)
$C_{z,0^+}$	fluid-phase concentration directly after entering column (moles m⁻³)
$C_{z,1^-}$	fluid-phase concentration before column exit (moles m⁻³)
$C_{z,1^+}$	fluid-phase concentration after column exit (moles m⁻³)
D_{micro}	effective adsorbed-phase microparticle diffusion coefficient of n-hexane (m² s⁻¹)
D_x	microparticle gas-phase diffusion coefficient of n-hexane (m² s⁻¹)
$D_{x,\infty}$	microparticle gas-phase diffusion coefficient of n-hexane at infinite temperature (m² s⁻¹)
D_m	molecular diffusion coefficient of n-hexane in hydrogen (m² s⁻¹)
D_y	macroparticle diffusion coefficient of n-hexane (m² s⁻¹)
D_z	fluid-phase diffusion coefficient of n-hexane (m² s⁻¹)
$D_{z,0^-}$	diffusion coefficient of n-hexane directly before column entrance (m² s⁻¹)
$D_{z,0^+}$	fluid-phase diffusion coefficient of n-hexane directly after column entrance (m² s⁻¹)

$D_{z,1^-}$	diffusion coefficient of n-hexane directly before column exit (m² s⁻¹)
$D_{z,1^+}$	fluid-phase diffusion coefficient of n-hexane directly after column exit (m² s⁻¹)
E_{act}	activation energy of D_{micro} (J mol⁻¹)
ΔH	heat of adsorption of n-hexane in the micropores (J mol⁻¹)
J	Jacobian of ODE system ( )
K_a	dimensionless adsorption equilibrium constant ( )
$K_{a,\infty}$	K_a at infinite temperature ( )
L_{bed}	column length (m)
L	number of fluid-phase spatial grid points ( )
M	number of macroparticle grid points ( )
N	number of microparticle grid points ( )
N_{Rx}	mass flux at microparticle boundary (moles m⁻² s⁻¹)
N_{Ry}	mass flux at macroparticle boundary (moles m⁻² s⁻¹)
Pe	Peclet number of the fluid phase ( )
R	gas constant (J mol⁻¹ K⁻¹)
Re	Re number of the fluid phase
R_x	microparticle radius (m)
R_y	macroparticle radius (m)
Sc	Schmidt number ( )
Sh	Sherwood number for the film ( )
T	absolute temperature (K)
T_{room}	room temperature (293 K)
k_f	film mass transfer coefficient (m s⁻¹)
t	time (s)
t^*	dimensionless time (t/τ)
v_{sup}	superficial carrier gas velocity (m s⁻¹)
v_{int}	interstitial carrier gas velocity (m s⁻¹)
x	microparticle radial coordinate (m)
y	macroparticle radial coordinate (m)
z	fluid-phase axial coordinate (m)
λ	Jacobian eigenvalue ( )
τ_y	macroparticle tortuosity ( )
τ_z	microparticle tortuosity ( )
τ	space time (L_{bed}/v_{sup}) (s)

Note: The following relationships exist between TEX-PEP definitions and those used in the Hayne's model.

$$D_c = \frac{D_x}{\varepsilon_x(1 + K_a)} \qquad D_{ax} = \frac{D_z}{\varepsilon_z} \qquad v_{int} = \frac{v_{sup}}{\varepsilon_x}$$

References

1 L. F. Gladden, P. Alexander, *Meas. Sci. Technol.* **1996**, 7, 423.

2 R. A. Williams, D. M. Scott (Eds.), *Frontiers in Industrial Process Tomography*, A. I. Ch. E. Publications, New York, **1995**.

3 R. A. Williams, D. J. Parker (Eds.), *Process Tomography: Principles, Techniques and Applications*, Butterworth-Heinemann Ltd., Oxford, **1995**.

4 E. Krestel (Ed.), Medical Imaging in Nuclear Medicine, Siemens Aktiengesellschaft, Berlin, 1990.

5 V. Lammers, *Light Collection and Scattering Corrections in the EUT PEP Detector*, M.Sc. Thesis, Eindhoven University of Technology, Eindhoven, The Netherlands, **1996**.

6 P. Hautojarvi (Ed.), *Positrons in Solids*, Springer-Verlag, Berlin, **1979**.

7 S. De Benedetti, C. E. Cown, W. R. Konnecker, H. Primakoff, *Phys. Rev.* **1950**, 77, 205.

8 R. Hakvoort, *Applications of Positron Emission Depth Profiling*, Ph.D. Thesis, Delft University of Technology, Delft, The Netherlands, **1993**.

9 M. A. M. Haast, *Performance of the EUT Positron Emission Profiling Detector*, M.Sc. Thesis, Eindhoven University of Technology, Eindhoven, The Netherlands, **1995**.

10 R. Miranda, R. Ochoa, W. F. Huang, *J. Mol. Catal.* **1993**, 78, 67.

11 J. Lahtinen, A. Verhanen, *Catal. Lett.* **1991**, 26, 67.

12 K. L. Cheng, Y. C. Jean, X. H. Liu, *Crit. Rev. Anal. Chem.* **1989**, 21, 209.

13 R. A. Ferrieri, A. P. Wolf, *J. Phys. Chem.* **1984**, 88, 2256.

14 R. A. Ferrieri, A. P. Wolf, *J. Phys. Chem.* **1984**, 88, 5456.

15 U. Baltensburger, M. Amman, U. K. Bochert, B. Eichier, H. W. Gjggeler, D. T. Jost, J. A. Kovacs, A. Tijder, U. W. Sherer, A. Baiker, *J. Phys. Chem.* **1993**, 97, 12325.

16 M. A. Dell, *J. Nucl. Med. Tech.* **1997**, 25, 12.

17 A. L. Feliu, *J. Chem. Ed.* **1988**, 655.

18 M. Dahlbom, L. Eriksson, K. Wienhard, in *Proceedings of the Nuclear Science Symposium and Medical Imaging Conference,* Norfolk, VA, USA, Oct. 30–Nov. 5, **1994**, IEEE New York, NY, USA, vol. 4, 1995, p. 1667.

19 C. J. Marriot, J. E. Cadorette, R. Lecomte, V. Scasnar, J. Rousseau, J. E. van Lier, *J. Nucl. Med.* **1994**, 35(8), 1390.

20 S. Siegel, S. R. Cherry, A. R. Ricci, Y. Shao, M. E. Phelps, in *Proceedings of the Nuclear Science Symposium and Medical Imaging Conference,* Norfolk, VA, USA, Oct. 30–Nov. 5, **1994**, IEEE New York, NY, USA, vol. 4, 1995, p. 1662.

21 P. S. Schewe, B. S. Stein, *Amer. Inst. Phys. Bull. Phys. News, May 16,* **1996**, 271.

22 M. R. Hawkesworth, D. J. Parker, P. Fowles, J. F. Crilly, N. L. Jefferies, G. Jonkers, *Nucl. Instr. Meth. Phys. Res.* **1991**, A310, 423.

23 F. Hensel, (Institut Sicherheitsforschung, Forschungszentrum Rossendorf, D-01314 Dresden, Germany). FZR, FZR-152, 12–16, **1996**.

24 M. R. Hawkesworth, D. J. Parker, in *Process Tomography: Principles, Techniques and Applications* (Eds.: R. A. Williams, D. J. Parker), Butterworth-Heinemann Ltd., Oxford, **1995**, p. 199.

25 C. J. Broadbent, J. Bridgwater, D. J. Parker, *Chem. Eng. J.* **1995**, 56, 119.

26 C. R. Bemrose, P. Fowles, M. R. Hawkesworth, M. A. O'Dwyer, *Nucl. Instr. Meth.* **1988**, A273, 874.

27 D. J. Parker, M. R. Hawkesworth, C. J. Broadbent, P. Fowles, T. D. Fryer, P. A. McNeil, *Nucl. Instr. Meth. Phys. Res.* **1994**, A348, 583.

28 D. J. Parker, A. E. Dijktra, T. W. Martin, J. P. K. Seville, *Chem. Eng. Sci.* **1997**, 52, 2011.

29 C. S. Stellema, J. Vlek, R. F. Mudde, J. J. M. de Goeij, C. M. van den Bleek, *Nucl. Instr. Meth. Phys. Res.* **1998**, A404, 334.

30 G. Jonkers, K. A. Vonkeman, S. W. A. van der Wal, R. A. van Santen, *Nature* **1992**, 355, 63.

31 G. Jonkers, K. A. Vonkeman, S. W. A. van der Waal, in *Precision Process Technology* (Eds.: M. P. C. Weijnen, A. A. H. Drinkenburg), Kluwer Academic Publishers, Dordrecht, The Netherlands, **1993**, p. 533.

32 A. V. G. Mangnus, L. J. van IJzendoorn, J. J. M. de Goeij, R. H. Cunningham, R. A. van Santen, M. J. A. de Voigt, *Nucl. Instr. Meth.* **1995**, B99, 649.

33 R. J. Jaszcak, K. L. Greer, C. E. Floyd, C. C. Harris, R. E. Coleman, *J. Nucl. Med.* **1984**, 25, 893.

34 S. Grootoonk, T. J. Spinks, T. Jones, C. Michel, *A. Bol, IEEE Med. Im. Conf. P.* **1992**, 1569.

35 R. F. Dannals, H. T. Ravert, A. A. Wilson, in *Nuclear Imaging in Drug Discovery, Development, and Approval* (Eds.: H. D. Burns, R. Gibson, R. Dannals, P. Siegel), Birkhauser, Boston, **1993**, p. 55.

36 A. P. Wolf, D. J. Schlyer, in *Nuclear Imaging in Drug Discovery, Development, and Approval* (Eds.: H. D. Burns, R. Gibson, R. Dannals, P. Siegel), Birkhauser, Boston, **1993**, p. 33.

37 R. H. Cunningham, A. V. G. Mangnus, J. van Grondelle, R. A. van Santen, *J. Molec. Catal.* **1996**, A107, 153.

38 T. Koerts, P. A. Leclercq, R. A. van Santen, *J. Am. Chem. Soc.* **1992**, 114, 7272.

39 T. J. McCarthy, C. S. Dence, S. W. Holmberg, J. Markham, D. P. Schuster, M. J. Welch, *Nucl. Med. Biol.* **1996**, 23, 773.

40 G. K. Mulholland, M. T. Vavrek, *J. Nucl. Med. Suppl.* **1994**, 35, 72.

41 S. C. van der Linde, W. P. A. Jansen, J. J. M. de Goeij, L. J. van IJzendoorn, F. Kapteijn, *Appl. Rad. Isot.* **2000**, 52, 77.

42 S. C. van der Linde, T. A. Nijhuis, F. H. M. Dekker, F. Kapteijn, J. A. Moulijn, *Appl. Catal. A* **1997**, 151, 27.

43 F. Helfferich, D. L. Peterson, *Science* **1963**, 142, 661.

44 S. H. Hyun, R. P. Danner, *AIChE. J.* **1985**, 31, 1077.

45 J. H. Hufton, R. P. Danner, *AIChE. J.* **1993**, 39, 962.

46 B. G. Anderson, F. J. M. M. de Gauw, N. J. Noordhoek, L. J. van IJzendoorn, R. A. van Santen, M. J. A. de Voigt, *Ind. Eng. Chem. Res.* **1998**, 37, 815.

47 N. J. Noordhoek, L. J. van IJzendoorn, B. G. Anderson, F. J. M. M. de Gauw, R. A. van Santen, M. J. A. de Voigt, *Ind. Eng. Chem. Res.* **1998**, 37, 825.

48 D. M. Ruthven, *Principles of Adsorption and Adsorption Processes*, Wiley Interscience, New York, **1984**.

49 D. M. Ruthven, C. B. Ching, in *"Modelling of Chromatographic Processes in Preparative and Production Scale Chromatography"* (Eds.: G. Ganetsos, P. E. Barker), Marcel Dekker, New York, **1992**.

50 J. Kärger, D. M. Ruthven, *Diffusion in Zeolites and Other Microporous Solids*, John Wiley & Sons, **1992**.

51 M. Suzuki, *Adsorption Engineering*, Kodansha, Tokyo, **1990**.

52 H. W. Haynes, *Chem. Eng. Sci.* **1975**, 30, 945.

53 E. N. Fuller, P. D. Schettler, J. C. Gidding, *Ind. Eng. Chem.* **1996**, 58(5), 19.

54 P. V. Danckwerts, *Chem. Eng. Sci.* **1953**, 1, 2.

55 G. F. Froment, K. B. Bischoff, *Chemical Reactor Analysis and Design*, 2nd Ed., John Wiley & Sons, New York, **1990**.

56 A. R. van Cauwenberghe, *Chem. Eng. Sci.* **1966**, 21, 203–205.

57 W. E. Schiesser, *The Numerical Method of Lines*, Academic Press Inc., New York, **1993**.

58 M. F. M. Post, *Stud. Surf. Sci. Catal.* **1991**, 58, 391.

59 S. L. Shannon, J. G. Goodwin Jr., *Chem. Rev.* **1995** 95, 677.

60 R. R. Schumacher, B. G. Anderson, N. J. Noordhoek, F. J. M. M. de Gauw, A. M. de Jong, M. J. A. de Voigt, R. A. van Santen, *Micropor. Mesopor. Mater.* **2000**, 35–36, 315.

61 J. Xiao, J. Wei, *Chem. Eng. Sci.* **1992**, 47, 1143.

62 C. L. Cavalcante, D. M. Ruthven, *Ind. Eng. Chem. Res.* **1995**, 34, 185.

63 O. P. Keipert, M. Baerns, *Chem. Eng. Sci.* **1998**, 53, 3623.

64 K. A. Vonkeman, *Exhaust Catalysis Studies Using In Situ Positron Emission*, Ph.D. Thesis, Eindhoven University of Technology, Eindhoven, The Netherlands, **1990**.

65 C. T. Campbell, G. Ertl, H. Kuipers, J. Segner, *J. Chem. Phys.* **1980**, 73(11), 5862.

66 V. A. Sobyanin, K. I. Zamaraev, *React. Kinet. Catal. Lett.* **1986**, 31(2), 273.

67 L. K. Verheij, J. Lux, A. B. Anton, B. Poelsma, G. Comsa, *Surf. Sci.* **1987**, 182, 390.

68 V. Matolin, E. Gillet, *Surf. Sci.* **1987**, 186, L541.

69 S. H. Oh, G. B. Fisher, J. E. Carpenter, D. W. Goodman, *J. Catal.* **1986**, 100, 360.

70 G. B. Fisher, S. H. Oh, J. E. Carpenter, C. L. DiMaggio, L. D. Schmidt, in *"Catalysis and Automotive Pollution Control"* (Eds.: A. Crucq, A. Frenet), Elsevier Science Publishers, Amsterdam, **1987**, p. 215.

71 S. B. Schwartz, G. B. Fisher, L. D. Schmidt, *J. Phys. Chem.* **1988**, 92, 389.

72 D. T. Lynch, G. Emig, S. E. Wanke, *J. Catal.* **1986**, 97, 456.

73 R. A. van Santen, B. G. Anderson, R. H. Cunningham, A. V. G. Mangnus, J. van Grondelle, L. J. van IJzendoorn, M. J. de Voigt, *Angew. Chem. Int. Ed. Engl.* **1996**, 35, 2785.

74 S. C. van der Linde, *Application of Positron Emission Profiling in Catalysis*, Ph.D. Thesis, Delft University of Technology, Delft, The Netherlands, **1999**.

75 N. J. Noordhoek, Ph.D. Thesis, Eindhoven University of Technology, Eindhoven, The Netherlands, **2000**.

76 A. C. M. van den Broek, J. van Grondelle, R. A. van Santen, *J. Catal.* **1999**, 185, 297.

77 Y. S. Matros, G. A. Bunimovich, *Catal. Rev. Sci. Eng.* **1996**, 38, 1.

9
TAP Reactor Studies

Olaf Hinrichsen, Andre C. van Veen, Horst W. Zanthoff and Martin Muhler

9.1
Introduction

9.1.1
What is the TAP Method?

Transient experiments are a powerful tool for gaining insight into the mechanisms of complex catalytic reactions and to derive rate constants for the individual steps involved [1]. However, kinetic information is only obtained when the transient change proceeds rapidly enough and the system cannot relax to equilibrium while being monitored [2]. A conventional nonsteady-state reactor set-up operating at atmospheric pressure provides a time resolution of only about one second, limited by the way in which the perturbation is created and by the inevitable broadening of sharp responses that occurs in a flow system. In contrast, the **T**emporal **A**nalysis of **P***roducts* reactor system allows fast transient experiments in the millisecond time regime with submillisecond signal sampling [3]. The enhancement in resolution of more than two orders of magnitude is achieved by using high-speed pulse valves and by establishing mass transport by diffusion without a carrier gas under high-vacuum conditions.

The construction of the TAP reactor system and the underlying experimental conditions are highly suitable for transient kinetic studies. First, the model used for the description of transport phenomena within the reactor should be as simple as possible to avoid complex mathematical descriptions. This is achieved by applying high-vacuum conditions, under which the mass transport may be expressed as Knudsen diffusion within a packed bed. The catalyst sample is therefore mounted as a fixed bed in a reactor which is dynamically evacuated at the outlet. Second, the shape of the perturbation to trigger the transient behavior should be well defined. In a typical TAP experiment, sharp and small pulses (width <2 ms,

10^{13}–10^{18} molecules per pulse) are used to study the interaction of reactants with a heterogeneous catalyst in the working state of the surface or to obtain detailed information on surface reactions. A broadening of the introduced pulse is avoided by minimizing the distance and the volume between the pulse valve and the reactor entrance. Thus, the input pulse of molecules to the reactor may be described as a Dirac delta function. Third, the analytical device should provide a high sensitivity and time resolution. Within the TAP set-up, a quadrupole mass spectrometer is used for fast quantitative analysis. As a benefit of the vacuum conditions, no capillary is needed, resulting in a very sensitive signal detection for TAP pulse responses with a high time resolution.

9.1.2
How Can We Classify the TAP Method?

TAP experiments can be regarded as a connecting link between conventional kinetic experiments with a packed bed of catalyst particles performed at atmospheric pressure and molecular beam studies with a catalytically active surface conducted under high-vacuum conditions. Despite the fact that the molecular beam technique may provide detailed mechanistic insights into a catalytic reaction by identifying desorbing intermediates, it suffers from the small number of collisions between the reactant molecules and the catalytic surface, which restricts its application to highly reactive surfaces. In contrast, kinetic experiments at atmospheric pressure can be performed with catalysts exposing surfaces with low reactivity. Unfortunately, these studies rarely provide information about reactive intermediates as the large number of intermolecular collisions in the gas phase often reduces their concentration to undetectably small amounts.

Typical parameters characterizing the working conditions of transient pulse techniques are summarized in Tab. 9.1. While molecular beam studies require

Tab. 9.1 Key features of transient pulse techniques. (a) The TAP-2 version allows an increased maximum sample weight and a smaller pulse size.

	molecular beam	TAP-1 reactor system	laboratory fixed-bed reactor
catalyst amount	–	10–500 mg[a]	10–10000 mg
pulse size	10^{14}–10^{19}	10^{13}–10^{18} [a]	10^{18}–10^{21}
gas–solid collisions	<10	100–10^{6}	$\gg 10^{6}$
gas–gas collisions	<10	<10^{3}	$\gg 10^{6}$
required reaction probability	10^{-7}–10^{-6}	10^{-18}–10^{-7}	10^{-20}–10^{-18}
catalyst contact time	–	1–5 ms	>1 s

flat surfaces, conventional powderous samples are typically examined in TAP experiments. Furthermore, the reaction probability of reactants within the catalytic fixed bed is much higher, allowing for pulse experiments and for the detection of intermediates with far less reactive catalysts [4–6]. The packed bed usually offers a large number of catalytic sites in comparison to the pulsed amount of reactant. In general, only a very small proportion of the surface of the catalyst is generally addressed by the reactant pulse and significant changes in the state of the surface are avoided. Thus, the TAP method enables the determination of reaction probabilities for catalysts with reproducible states of the exposed surfaces, which can be changed in a controlled way by additional pretreatment procedures.

9.1.3
What is the TAP Method for?

The first review paper about TAP reactor studies was published in 1988 [3]. Since then, research groups both in academia and industry throughout the world have studied heterogeneously catalyzed reactions by this fast transient technique. The main research topics can be summarized as follows:

- testing catalysts for which the state of the surface is well-defined, e.g. to determine activity and selectivity,
- elucidation of complex catalytic processes, for instance distinction between parallel and consecutive reaction pathways, determination of the nature of the active catalyst during the reaction, e.g. by means of transient response experiments (temperature-programmed experiments, step transient experiments, isotopic switching experiments, etc.),
- fundamental model studies aiming at a mechanistic understanding in terms of microkinetic models,
- assistance in the design of new catalysts.

9.2
Description and Operation of the TAP Reactor System

9.2.1
The TAP Reactor System

Almost two decades ago, John T. Gleaves and co-workers at Monsanto developed an apparatus for carrying out fast transient experiments, the so-called temporal analysis of products (TAP) reactor system. To date, approximately 20 set-ups of the TAP-1 reactor system and the more modern TAP-2 version have been built. The original set-up (TAP-1 version) was presented in 1986 by Gleaves and co-workers [7] and is described in various patents [4, 8, 9]. A simplified schematic overview of

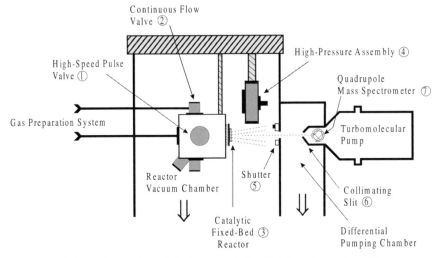

Fig. 9.1 Schematic overview of the key parts in the TAP-1 reactor system [9].

the key parts in the TAP-1 reactor system is shown in Fig. 9.1. The TAP-1 set-up consists of a gas preparation system, three horizontally mounted differentially pumped, interconnecting vacuum chambers housing the dosing valves, a catalytic fixed-bed reactor system, and a mass spectrometer as well as data acquisition/control facilities.

The gas preparation allows the volumetric controlled preparation of gas mixtures in two separate blend tanks. These blends may be binary mixtures of a reactant gas with an internal standard component for calibration or multicomponent mixtures for carrying out catalytic experiments. Studies with condensed substances are conducted by using the vapor phase, provided that the resulting pressure is sufficiently high. The minimum vapor pressure at a given gas preparation temperature is approximately 20 mbar. In addition, gas mixtures can be introduced through an external port. The supply of gases for each valve located in the reactor vacuum chamber is accomplished by separate gas lines. The dosed amounts are calibrated by recording the pressure drop within the known volume of the feed system during supplying a large train of pulses. In addition to the two high-speed pulse valves ①, one or two solenoid valves ② are installed to allow experiments with a continuous gas feed. The valves are directly connected to a temperature controlled "zero volume" manifold, which represents the interface to the reactor ③.

The conventional catalytic fixed-bed reactor of the commercial TAP-1 set-up has been described by Svoboda et al. [10]. Using highly active reactants is complicated by signal broadening or by irreversible adsorption. Avoiding interactions and reactions with the oxidized steel of the reactor requires special care. A good example of the need for modified constructions is the use of a quartz reactor during experiments on the selective oxidation of methanol reported by Lafyatis et al. [11]. A detailed description of a more versatile reactor design has recently been presented by van Veen et al. [12]. A schematic illustration of this catalytic TAP-1 fixed-bed reactor is shown in Fig. 9.2. However, even when using improved reactor designs, blind experiments are necessary to ensure the absence of experimental artefacts, especially when using catalyst samples with only low-surface areas. The temperature of the reactor within a surrounding furnace is measured by thermocouples and can be adjusted in the range of 300 to 1200 K by a PID temperature controller, which also allows temperature-programmed experiments to be carried out.

The reactor vacuum chamber can be separated from the remaining vacuum system by closing an adjustable shutter system ⑤. This avoids breaking the ultra-high vacuum in those parts housing the analytical device while the reactor chamber is vented for changing the reactor or for maintenance of the dosing system. The time required for changing the reactor and re-establishing the necessary vacuum conditions is further decreased by the use of very large pumping capacities. A 10″ oil diffusion pump is mounted below the reactor chamber, connected via a gate valve for closing off the pump during maintenance of the reactor.

The differential chamber is pumped by a 6″ oil diffusion pump and terminated by a collimating slit ⑥, ensuring that molecules leaving the reactor only pass through the chamber in an essentially straight path to reach the mass spectrometer for detection. Thus, the detected signal is not influenced by delayed or scattered molecules and yields a true response of the molecular flux leaving the reactor. The

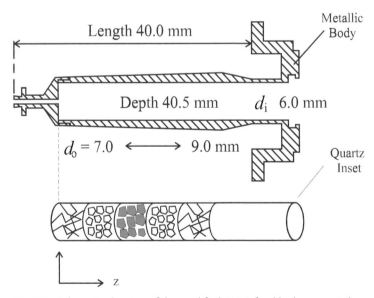

Fig. 9.2 Schematic drawing of the modified TAP-1 fixed-bed reactor (taken from [12]).

sampling of data is performed by using a UTI 100C **Q**uadrupole **M**ass **S**pectrometer ⑦ located in the detector vacuum chamber, which is continuously evacuated by a trubomolecular pump. During a pulse experiment, the response curve of a fixed characteristic **A**tomic **M**ass **U**nit is recorded with a submillisecond time resolution. As only a very small portion of the catalyst surface is addressed by a typical reactant pulse, the state of the surface is not changed significantly. Therefore, a signal with an enhanced signal-to-noise ratio can be obtained as an average of several pulse responses. For the same reason, it is possible to record the response of different components for the same state of the surface by subsequent experiments with other AMUs in a similar way. Obtaining the pulse responses as an average of several experiments requires that the opening of the valves is strictly correlated with the time scale of the data acquisition. Thus, the control of the experiment and the data aquisition is performed by a computer. The electronic triggering also allows operation of the pulse valves with a defined offset time, which may be used to dose different reactants independently. To obtain information about the complete product spectrum of a catalytic reaction, it is also possible to record scans in a wide AMU range with a much lower time resolution.

A modified TAP-1 set-up enables operation of the reactor in the "high-pressure" mode [13, 14]. This mode may be used during a pretreatment of the catalyst or a catalytic experiment in a continuous flow of carrier or reactant gases at ambient pressures. It is achieved by sealing the exit of the reactor by the "**H**igh-**P**ressure **A**ssembly" ④, allowing for operation up to 3 bar. The analysis of the product gas can be performed with the QMS by directing a small portion of the effluent through a pinhole leak towards the mass spectrometer [15]. The majority of the product gas is vented through an external bleed valve, where an additional analysis may be carried out.

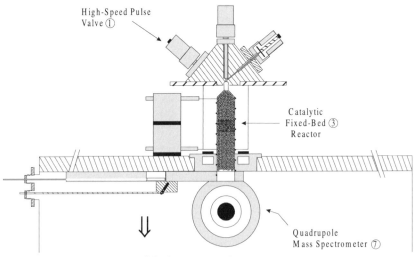

High-Speed Pulse
Valve ①

Catalytic
Fixed-Bed ③
Reactor

Quadrupole
Mass Spectrometer ⑦

Fig. 9.3 Schematic overview of the key parts in the TAP-2 version [16].

A second revision of the set-up was introduced later and is called TAP-2 (Fig. 9.3, taken from [16]). The main difference from the original set-up is the location of the fixed-bed reactor and the dosing valves within the molecular construction, allowing a simpler design of the vacuum system. These components are accessible from the exterior of the TAP-2 set-up and only the inner volume of the reactor is evacuated. This modular construction offers a better heat transfer between the reactor heating and the reactor wall and allows easier maintenance of the valves. Furthermore, different types of reactors (varying in length, inner diameter, material, etc.) can be used and, due to the small volume to be pumped off, only one oil diffusion pump is necessary to handle the whole system. Additionally, the mass spectrometer is located very close to the reactor outlet resulting in an enhanced sensitivity.

Finally, the multitrack set-up [17] can be regarded as a "TAP-like" system. In contrast to the conventional set-up, the simultaneous use of several mass spectrometers allows for the parallel real-time detection of different AMUs.

9.2.2
Types of Experiments

Depending on the kind of information required, different experimental modes are used. These modes may be roughly divided into pulse experiments and investigations using a continuous gas feed (for an overview see Table 2).

Single-pulse experiments to derive diffusivities (D_i^{eff}) and heats of adsorption (ΔH_{ads})

Pulse experiments can be applied to determine the effective Knudsen diffusion coefficient $D_{K,i}^{eff}$ wthin fixed beds. For this purpose, the average residence time $t_{a,i}$ for small, single pulses of non-adsorbing gases (less than 2×10^{15} molecules) within the reactor is measured. Eqn. 1 allows the calculation of $D_{K,i}^{eff}$ within an isothermal fixed bed with a given length L.

$$D_{K,i}^{eff} = \frac{L^2}{2\, t_{a,i}} . \tag{1}$$

The applied pulse size has to be restricted to ensure that mass transfer only occurs by Knudsen diffusion (see section 9.4.2.1).

Another important use of single-pulse experiments is the characterization of interactions between adsorbing molecules and the catalyst. A quantification of the obtained results in terms of the heat of adsorption is possible using expression 2 derived from the analytical model for simple adsorption processes by Gleaves et al. [3]. The measured average residence time $t_{a,j}$ of the adsorbing molecule j is compared to the theoretical average residence time $t_{a,i,j}$ for the same molecule without interaction with the catalyst surface, which is calculated from the average resi-

Tab. 9.2 Overview of the experimental modes and main information obtained.

Experiment	Obtained Information	Selected Refs.
Single-pulse technique	Diffusivities	18, 19
	Heats of adsorption	3, 5, 20
	Product formation sequence	3, 5, 21
	Number of adsorption sites	40
	Elucidation of reaction pathways	15
Sequentialpulse technique	Lifetime of surface intermediates	this work
	Elucidation of reaction pathways	15
Flow experiment	Role of catalyst bulk-dissolved and lattice species in the reaction	22, 23
Temperature-programmed desorption (TPD), reaction (TPSR)	Thermal stability of intermediates	24
	Redox properties of solid oxide catalysts	35

dence time of a non-adsorbing reference gas $t_{a,i}$ using Eqn. 3. Thus, the change in the average residence time due to the mass dependency of the diffusion velocity of molecules with different molar masses M is taken into account.

$$ln\left(\frac{t_{a,j}}{t_{a,i,j}} - 1\right) = \frac{-\Delta H_{ads,j}}{RT} + ln\,K = \frac{E_{des,j} - E_{ads,j}}{RT} + ln\,K. \tag{2}$$

$$t_{a,i,j} = t_{a,i}\sqrt{\frac{M_i}{M_j}}, \tag{3}$$

where K is defined as the ratio of the pre-exponential factors of adsorption and desorption. Therefore, the heat of a completely reversible molecular adsorption can be derived from the slope of the straight line obtained by plotting $ln\left(\frac{t_{a,j}}{t_{a,i,j}} - 1\right)$ versus T^{-1}. However, the model used enforces some restrictions upon the application of this method. First, the adsorption process must be completely reversible within the investigated temperature range, as can be verified by complete detection of the pulsed amount of reactants within the response. Therefore, the lower temperature limit is determined by the increase in time and the decrease in signal intensity necessary to ensure a complete detection of the pulsed molecules. An upper temperature limit often arises from the occurrence of chemical reactions, which, in turn, change the pulse response. Second, the pulse size used has to be limited to ensure that the mass transport only occurs by Knudsen diffusion.

Input Pulses:

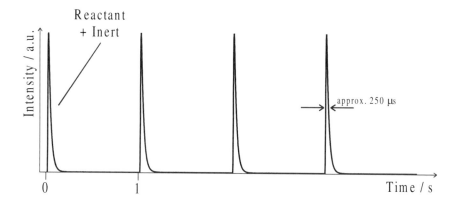

Pulse Responses:

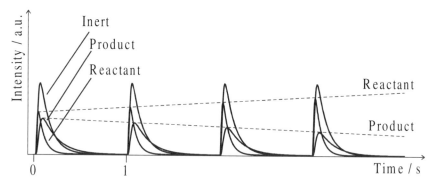

Fig. 9.4 Key features of a TAP single-pulse experiment.

Series of pulse experiments from a single valve

Finally, the results obtained by another application of single-pulse experiments are illustrated schematically in Fig. 9.4. The catalyst is adjusted to a reproducible state by appropriate pretreatments. Examples of these pretreatments are an oxidation of the catalyst or the adsorption of one reactant. Information on the catalytic perform-ance is obtained by pulsing a mixture of a reactant and a non-reacting reference gas at different temperatures and detecting the pulse responses for all components leaving the reactor. The integral size of the pulse responses allows calculation of conversions and yields, whereas the constant size and shape of the reference gas response peak ensures that the dosed amount of reactant does not change during the experiment. Due to the high time resolution, information on the sequence of product formation can often be obtained by a comparison of the normalized pulse responses for different products. The product formation on the surface requires a certain period of time for each step in the reaction sequence. Therefore, the pulse responses of secondary products are delayed in comparison to those of primary products, while components formed during the same reaction step appear simultaneously. However, the former statement is only valid if the time for the diffusive mass transport is (much) smaller than the time required for reaction and if the desorption of products into the gas phase is a fast process and if the surface reactions are virtually irreversible.

Furthermore, a large number of pulses (so-called multipulses) can be used to change the surface state of the catalyst in a distinct way, e.g. the surface reduction of an oxide catalyst when pulsing a reactive hydrocarbon. Therefore, these experiments may also probe the performance of the catalyst under different working conditions. In contrast to the first applications, these pulse experiments are not restricted to Knudsen diffusion conditions and allow the use of large pulses. However, in general, the pulsed amount of reactant is very small compared to the number of catalytic sites and the change in the product formation shown in Fig. 9.4 is strongly exaggerated.

Series of sequential-pulse experiments from separate valves (two reactants)

Sequential-pulse experiments are another experimental mode, in which two reactants are separately fed to the catalyst by triggering the pulse valves with a defined offset time. The inputs and transient responses obtained by this type of experiment are illustrated schematically in Fig. 9.5. The first reactant is adsorbed on the catalyst during the first pulse and forms an intermediate. When the second reactant is fed into the reactor, chemical reaction occurs and product formation is observed. Information about the lifetime and the amount of the surface intermediate can be derived from the product yields as a function of the offset time between the pulses of the two reactants and their pulse size.

Experiments with continuous viscous gas flow at low pressure (<10 mbar)

Experiments with a continous gas feed allow the use of larger amounts of reactants. Therefore, more pronounced changes in the state of the catalyst occur and can be traced by virtue of the high time resolution and comparably low reaction rate under reduced pressure conditions by recording the changes in the product spectra as a function of reaction time. In general, the same information can be obtained using conventional flow experiments at ambient pressure. The time window for the investigated processes addressed, however, is shifted to longer durations, which in turn, allows the tracing of these processes at a reasonable extent.

Experiments with continous viscous gas flow at high pressure (1 to 3 bar)

The pretreatment of a catalyst often requires the use of large amounts of reactants or is restricted to partial pressures which are not applicable under vacuum opera-

Input Pulses:

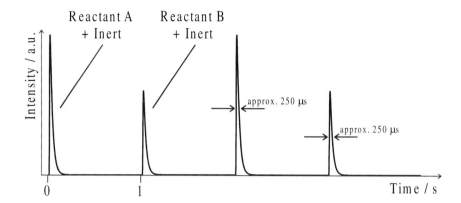

Pulse Responses:

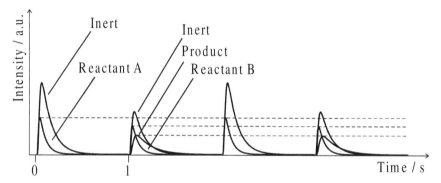

Fig. 9.5 Key features of a TAP sequential-pulse experiment.

tion. Furthermore, *in situ* activation of a catalyst under atmospheric pressure within the TAP reactor may be required to obtain relevant information on catalysts with the surface in the working state. Therefore, the TAP system provides the possiblity of pressurizing the reactor by closing the exit with a high-pressure assembly. A good example of the value of the "high-pressure" mode is given in the work by Rigas et al. [25]. The activation of a silver catalyst for ethylene epoxidation was studied by continuously recording the composition in the reactor effluent during pretreatment runs with the high-pressure mode applied and subsequently performing pulse experiments under evacuated conditions.

Temperature-programmed experiments
Up to now, the TAP set-ups have only rarely been used for these types of experiments. In general, all possible applications, such as TPD, TPSR, TPO (oxidation),

or TPR (reduction) can be performed. Results of TPD experiments allow the characterization of adsorbates on the catalyst. In addition to pulse experiments, which probe the adsorption of reactants, these studies provide detailed information on the thermal stability of adsorbed intermediates. One example of this approach is the study by Dewaele and Froment on the adsorption of CO and CO_2 on alumina [24].

Experiments with isotopes

The use of expensive isotopes is especially helpful in mechanistic studies, as the pathway leading to product formation can be traced by the different incorporation within the final products. A good example is the clarification of the role of bulk-dissolved or lattice atoms in the Mars–van Krevelen mechanism. In this case, the catalyst was pretreated using one isotope of the reactant under investigation, e.g. $^{16}O_2$, and the experiments were carried out with an educt containing another isotope, e.g. $^{18}O_2$. The time dependence of the isotopic distribution within the products allowed insights into the reaction mechanism. Furthermore, the use of isotopes often helps to simplify the identification of products in a mass spectrum. For example, N_2O and CO_2 cannot be quantitatively separated using the usual ^{14}N, ^{12}C, and ^{16}O isotopes only, because both compounds exhibit fragments in the mass spectrum at AMU 44, 28, and 16. The use of ^{15}N or ^{13}C as the source of nitrogen or carbon increases the mass of one of these components by unity and therefore deconvolution is no longer necessary.

9.3
Modeling the TAP Experiment

The mathematical framework for modeling TAP pulse experiments is based on the solution of the time-dependent mass, energy, and momentum balance equations with the knowledge of the estimated kinetic parameters in the model (e.g. rate constants, which can be further split up into the activation energy E_i and the pre-exponential factor A_i using the Arrhenius' law). In general, the modeling procedure comprises the following steps:

1. Formulation of the mass, energy, and momentum balance equations using an appropriate reactor transport model.
2. Choosing initial and boundary conditions to describe the underlying experimental conditions (e.g. step function experiment, pulse experiment, isotopic switching experiments).
3. Implementing a model that adequately describes the reaction mechanism and the surface chemistry (e.g. Arrhenius' law, Langmuir–Hinshelwood (LH) mechanism, kinetic Monte Carlo (KMC) simulation).
4. Performing a calculation for the model based on the reaction conditions used for the TAP pulses.
5. Optimization routine (advanced mode, optional).

The degree of mathematical complexity for the solution will become high due to the requirement to describe the involved processes by nonlinear partial differential equations (PDEs), which have to be solved concomitantly with the underlying initial and boundary conditions. In the case of steady-state TAP flow experiments, the differential equations are reduced to ordinary differential equations (ODEs). In the advanced mode of TAP modeling, an optimization routine can be applied in order to extract kinetic parameters directly from the pulse response data. Hence, an appropriate objective function has to be chosen for a sensitivity analysis of the dependent variables. Generally, using least squares, which is defined as the weighted sum of the squares of the deviations in the observations of all events from the predictions of the postulated model, is appropriate for single-response data. The more complex multi-response estimation is preferred for those cases when there are two or more responses with unknown relative precisions and/or possibly interdependent error distributions. In general, robust optimization methodologies like the Nelder–Mead algorithm or the Generalized Reduced Gradient algorithm are often used optimize the adjustable parameters.

9.3.1
The Basis of TAP Pulse Modeling – Description of Gas Transport

In general, gas transport through the reactor proceeds either by inertia-dominated flow, viscous flow, Knudsen flow, or a combination of the latter two. Since small pulse sizes can be applied in experimental TAP studies, pulse response experiments are performed in the Knudsen regime. However, inertial effects at the onset of a gas pulse for very short times or viscous flow during the opening time of the pulse cannot be excluded, and these should be minimized when a detailed quantitative analysis of the pulse response curves is pursued. The ideal pulse size can be estimated from the kinetic theory of gases assuming that the mean free path must be larger than the distance between catalyst particles. As a rule of thumb, a number of molecules as large as 2×10^{15} per pulse can be considered as the highest limit for purely Knudsen diffusion. The validity of this assumption can be demonstrated experimentally. The onset of the Knudsen flow regime is easily determined by lowering the pulse intensity. Thereafter Knudsen diffusion is maintained as long as the experimental TAP results show that the normalized pulse shape for the diffusion–only case is independent of the pulse intensity.

In the Knudsen flow regime, the equation of continuity for a non-reacting gaseous species i is given by the one-dimensional unsteady-state diffusion equation (Fick's second law)

$$\varepsilon_b \frac{\partial c_i}{\partial t} = D_{K,i}^{eff} \cdot \frac{\partial^2 c_i}{\partial x^2} , \qquad (4)$$

where c_i is the gas-phase concentration, t the time, and x the reactor coordinate. $D_{K,i}^{eff}$ represents the effective Knudsen diffusion coefficient, whereas the bed void fraction ε_b is included according to the relationship

$$D_{K,i}^{eff} = \frac{d_c}{3} \frac{\varepsilon_b}{\tau} \sqrt{\frac{8\,RT}{\pi M_i}}, \tag{5}$$

where d_c is the average interpaticle distance, τ the tortuosity, R the ideal gas constant, T the temperature, and M_i the molecular weight of species i. The Knudsen diffusivities for different components at different temperatures can be related to each other using the formula

$$D_K^{eff} = D_{K,i}^{eff} \cdot \sqrt{\frac{T}{T_i} \cdot \frac{M_i}{M}}, \tag{6}$$

where the value for the inert gas, $D_{K,i}^{eff}$, with a molecular mass of the inert species M_i is measured experimentally at the temperature T_i. Again, the validity of the Knudsen diffusion model can be checked by plotting the square root of the molecular weight as a function of the average residence time. In the case of a purely diffusive flow the plot should yield a straight line. For the TAP-1 reactor system this linear correlation is indeed obtained, except for hydrogen and helium if their pulse intensities are close to the upper limit of 2×10^{15} molecules/pulse. This is due to the fact that these gases are pumped off by the oil diffusion pumps less rapidly than other gases having larger molar masses and therefore remain in the detection chamber to a larger extent.

9.3.2
Analytical Solution for TAP Pulse Experiments

An analytical solution can be derived as long as the underlying differential equations remain linear. Based on several assumptions, i.e. that the pulse can be described by a Dirac delta function, results for irreversible or reversible molecular adsorption as well as for a simple lumped process, which consists of molecular adsorption and desorption and first-order reaction, have been derived by Gleaves et al. [3]. The solution for these simple cases is straightforward. The governing PEDs can be solved using either the method of separation of variables [3] followed by a momentum analysis, or the Laplace transform with respect to time followed by analytical integration of the resulting set of ordinary differential equations in the Laplace domain [15, 26]. Subsequently, a fast Fourier transformation (FFT) can be applied in order to analyze the kinetic parameters. In general, the analysis of the zeroth, first, and second moments of the experimental curves can give useful insights into the interaction of the pulsed molecules with the inert material or the catalyst. The basic theory of the momentum-based solution for TAP pulse experiments was introduced by Gleaves et al. [3] and has recently been further improved [27]. According to this basic theory, the zeroth momentum can be used to calculate variables such as the conversion, selectivity, or yield, the first momentum is related to the average residence time for a molecule within the TAP fixed-bed reactor, while the higher momenta provide more information about the interaction kinetics. Application of the momentum analysis is presented in Section 9.4.2.

However, extensions to more complex problems, which involve dissociative adsorption or associative desorption, bi- or higher molecular surface reactions, higher fractional coverages of the adsorbed species, intraparticle diffusion, or initial coverage of preadsorbed species, render the equations nonlinear, and hence they have to be solved by means of numerical methods (Section 9.3.3.).

9.3.3
Numerical Solution for TAP Pulse Experiments

A more complex reaction mechanism results in a set of partial differential equations, which has to be solved without any *a priori* assumptions regarding possible rate-determining steps. Accordingly, a simple adjustment to other types of models including multiple adsorption sites, different types of adsorption, i.e. Langmuir, Freundlich, or Temkin isotherm, as well as complex reaction networks can be easily included in the underlying mathematical model. A numerical solution technique, predominantly the method of lines, is applied resulting in a set of ordinary equations, which can be solved using common integration routines.

Tab. 9.3 Overview of reactor models, solution of the partial differential equations (PDEs), and parameter fitting procedures.

Reactor Model
– One-zone model with a Dirac delta function as boundary conditions (b.c.) [3, 15, 27]
– One-zone model with experimentally determined b.c. [10]
– Two-zone model with an inert ideally stirred fixed-bed zone [17, 26]
– Three-zone model using a Dirac delta function as b.c. [27, 28]
– Three-zone model using a linearly decreasing flux at the inlet as b.c. [29]
– Three-zone model using a linearly decreasing input function as b.c. [29, 30]
– Thin-zone model [31]

Solution Method of the Differential Equations
– Analytical solution using the method of separation of variables [3, 27]
– Laplace transform [26] followed by inverse fast Fourier transform [10, 15]
– Numerical solution [17, 27–30]

Fitting Procedure
– Moment analysis [3]
– Least-square fitting [15, 10, 17, 28–30]

9.3.4

Reactor Models

Several reactor transport models have been developed in order to suitably describe the transport characteristics in a TAP experiment (Tab. 9.3). Two typical types of TAP fixed-bed reactors are often used: the so-called 'one-zone' reactor with a mixture of inert material and catalyst, which was introduced by Gleaves et al. [3], and the most commonly used reactor in experimental TAP studies called 'three-zone' reactor, in which the catalyst zone is sandwiched between two beds of inert material.

9.3.4.1 **One-zone reactor model**

The following assumptions for the one-zone reactor model are applied:

- The catalyst bed is uniformly packed, i.e. the fractional voidage in the bed is constant.
- There exists no radial gradient of concentration in the catalyst bed.
- Axial or radial temperature gradients are neglected.
- The diffusivity of each gaseous species i is kept constant resulting in an effective Knudsen diffusion coefficient $D_{K,i}^{eff}$.

The one-dimensional continuity equations in the catalytic bed for the gaseous components and adsorbed species are given by

$$\varepsilon_b \frac{\partial c_i}{\partial t} = D_{K,i}^{eff} \cdot \frac{\partial^2 c_i}{\partial x^2} + \rho_b \cdot \sum_{j=1} r_{ij} \quad \text{for gaseous species} \tag{7}$$

$$\frac{\partial \Theta_i}{\partial t} = \frac{1}{N_z} \sum_{j=1} r_{ij} \quad \text{for surface intermediates,} \tag{8}$$

where the rate expression for formation of species i in reaction step j is given by

$$r_{ij} = v_{ij} \cdot r_j = k_j \cdot \prod_l c_l^{n_l} \prod_m \Theta_m^{n_m} . \tag{9}$$

Here, ρ_b is the catalyst bed density (kg_{cat}/m^3), N_z the total number of sites per catalyst weight (mol/kg_{cat}), n the reaction order (dimensionless), Θ_i the fractional coverage with surface species i (dimensionless). The stoichiometric coefficient of component i in reaction j is represented by v_{ij} (dimensionless), while r_{ij} is expressed in terms of $mol/(kg_{cat} \cdot s)$.

Example of a simple adsorption–desorption–diffusion case

When a pulse containing a single reactant A is traveling through the catalytic fixed-bed, adsorption on the catalyst surface, reaction (e.g. isomerization), and desorption take place. It is assumed that the first-order kinetics of Langmuirian adsorption can be described by the following mass balance for the adsorbed species A^{-*}:

$$\frac{\partial \Theta_A}{\partial t} = k_{ads} \cdot c_A \cdot \Theta_* - k_{des} \cdot \Theta_A . \tag{10}$$

To simplify matters, the rate constants for the elementary steps of adsorption and desorption and the site concentration are in general lumped into new rate constants termed k_{ads} and k_{des}, which, are typically expressed in terms of $m^3/(mol \cdot s)$ and s^{-1}, respectively. The following dimensionless mathematical model can be used to describe the adsorption–desorption–diffusion case in TAP pulse response experiments performed in the Knudsen regime:

$$\frac{\partial C(t,X)}{\partial t} = \frac{\partial^2 C}{\partial X^2} - k_{ads}(X) \cdot C(t,X) \ (1 - \Theta(t,X)) + k_{des} \cdot \Theta(t,X) \tag{11}$$

$$\frac{\partial \Theta(t,X)}{\partial t} = k_{ads}(X) \cdot C(t,X) \ (1 - \Theta(t,X)) - k_{des} \cdot \Theta(t,X) , \tag{12}$$

where t, C, X, k_{ads}, and k_{des} are defined in a dimensionless form. The boundary condition (b.c.) at the reactor entrance ($X = 0$) specifies that there is no flux when the valve is closed ($t = 0$). The b.c. at the exit of the reactor ($X = 1$) states that the reactor is maintained under vacuum conditions. Hence, the gas-phase concentration of the species i in the flux leaving the reactor is, in a good approximation, close to zero. During the opening of the valve, the pulse can be described by a Dirac delta function, assuming that the inlet pulse width is negligibly small. Therefore, the initial and boundary conditions are expressed in dimensionless form as follows:

$$C(0,X) = 0 \quad \Theta(0,X) = 0 \tag{13}$$

$$\frac{\partial C(t,X)}{\partial x}\bigg|_{X=0} = -\delta(t)_X \quad C(t,1) = 0 , \tag{14}$$

where $\delta(t)_X$ is the Dirac delta function. However, the boundary conditions for the application of the equation to TAP vacuum pulse experiments have been the subject of some controversy in the literature [3,

15, 29]. Modifications to Eqn. 14 have been introduced by Rothaemel and Baerns [29] in order to describe the performance of the valve more closely to the 'real' behavior. Instead of a Dirac input pulse, the b.c. at $X = 0$ was varied linearly with t within the opening time of the valve. Based on the assumption that the number of active sites is much higher than the number of molecules, i.e. the sum $1 - \Theta(t, X)$ in Eqn. 12 can be set equal to 1, Fig. 9.6 shows how the shape and the magnitude of the response curves significantly change when the ratio of the adsorption and the desorption rate constants is varied.

On the basis of an exit flow analysis, Gleaves et al. [27] established fingerprints for simple adsorption and reaction kinetics, which they derived from analytical solutions. These allow a distinction to be made between irreversible adsorption/reaction, reversible adsorption, fast adsorption and slow desorption, as well as fast adsorption and desorption.

9.3.4.2 Three-zone reactor model

The more complex, so-called 'three-zone-model' is used to describe the hydrodynamics in a reactor packed with layers of inert/catalyst/inert. The set of partial differential equations has to be solved, whereas in the inert zone the terms for adsorption and desorption are ignored. Additional boundary conditions are required at the boundaries between the three sections.

9.3.5
What is the Best Model?

Let us compare the merits of the different reactor models.

1. Recently, a two-dimensional model taking into account diffusion in both radial and axial directions was proposed in order to describe the characteristics of the symmetrical cylindrical fixed-bed reactor more precisely [32]. The results of theoretical studies showed that in this case the use of the one-dimensional model was justified.

2. Shekhtman et al. [31] developed a new reactor model for TAP pulse response experiments called a 'thin-zone reactor', whereby diffusion and chemical reaction can be separated. However, the applicability of this reactor concept for unraveling other than very fast reactions is questionable.

3. The degree of modeling in the advanced mode can be quite high, especially when aiming at an optimal agreement of the calculated values with the experimental data. However, it should be noted that the complexity of the underlying model is justified compared to the usefulness of the kinetic information obtained.

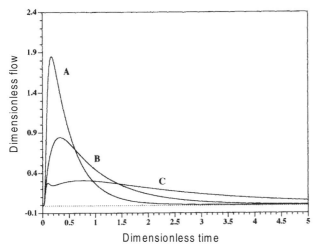

Fig. 9.6 Comparison of examples for the diffusion-only case with the reversible adsorption-diffusion case: (A) $k_{ads} = 0$; (B) $k_{ads} = 20$, $k_{des} = 20$; (C) $k_{ads} = 20$, $k_{des} = 5$. Adapted from [27].

9.4
Selected Applications in Heterogeneous Catalysis

9.4.1
Historical Overview

In this section examples of the application of the different experimental TAP methods in practice to a selected number of investigations are explained and evaluated for their relevance to technical conditions. Since no comprehensive review of the application of the TAP reactor system has yet appeared in the literature, the following listing of research topics should be regarded as an overview of TAP studies, which includes previous results of different research groups. Early detailed studies dealt with ethylene epoxidation over silver powder [25]. Much attention on unraveling complex reaction networks has been paid to the oxidation of propylene to acrolein [15], the partial oxidation of methane [33], *n*-butane [20], and propane [34], the ammoxidation of propane [35], the oxidation of CO over Pt [36], as well as the oxidation of alkylaromatics over vanadia/titania catalysts [21, 37]. Recently, a TAP study of *iso*-octane cracking over commercial fluidized catalytic cracking (FCC) catalysts revealed the ability of this method to determine the key mass transport and reaction kinetic parameters that control the overall process [18]. The use of strongly adsorbing reactants, e.g. for the selective oxidation of methanol to formaldehyde on an industrial Fe/Cr/Mo catalyst [11] and low-surface silver [12], or high-temperature usage [38] demonstrate the experimental need for an improvement of TAP fixed-bed reactors. Furthermore, the TAP studies on the adsorption of silanes on polycrystalline silicon [39] as well as the pyrolysis chemistry and kinetics of vapor-phase epitaxy (VPE) and chemical vapor deposition (CVD) organometallic precursors serve as nice examples of the applicability of this fast transient method.

9.4.2
Investigation of Non-reactive Interactions of Gases with Solid Catalysts

For kinetic purposes, the adsorption and desorption properties of gases on solid catalysts are important parameters to determine. The TAP pulse response method is applicable to irreversible as well as reversible interactions, as is explained in the following.

9.4.2.1 Diffusion coefficients

Analysis of TAP pulse response leads to the determination of the effective diffusion coefficient D_i^{eff}, which is an important value in further simulations to obtain kinetic parameters (activation energies and frequency factors, adsorption enthalpies, etc.). For pulse sizes below 2×10^{15} molecules/pulse, this is usually identical to the Knudsen diffusion coefficient ($D_{i,K}^{eff}$), but may differ for larger pulses due to molecular diffusion or viscous flow contributions. While these effective diffusion coefficients are necessary to determine other constants using the same catalyst bed, they can hardly be used for other purposes. On the contrary, inner particle diffusion coefficients have recently been calculated from TAP pulse experiments for zeolitic material [18, 19], which are in excellent agreement with diffusivities determined by other microscopic techniques, e.g. pulse-field-gradient NMR.

9.4.2.2 Irreversible interaction

For irreversible processes, the pulsed number of molecules entering the reactor diminishes over the solid and thus the area of the response signal *Area* detected by the mass spectrometer is lowered accordingly. With increasing number of pulses, the area under the response curve $Area_i$ will increase up to a saturation limit $Area_{sat}$ where no more gas molecules interact irreversibly with the catalyst because all adsorption sites are occupied. The number of adsorption sites then can be calculated from:

$$\frac{n_{ads}}{m_{cat}} = \sum_i (Area_{sat} - Area_i) \cdot S \quad [\text{sites/g}] \tag{15}$$

with S being the calibration factor for the gaseous compound. Therefore, the pulse mode can be used to titrate the number of adsorption sites (cf. Fig. 9.7). The method is, however, restricted to interactions where no disorption process occurs in parallel on a larger time scale than detectable in the TAP reactor

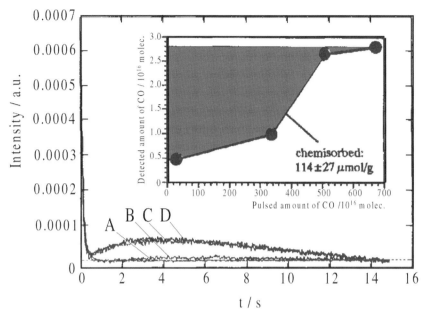

Fig. 9.7 CO response signals (2.8 × 10^{16} molecules CO/pulse) from a Pd(10 wt-%)/Al$_2$O$_3$ catalyst after different amounts of pulsed CO: A: 2.8 × 10^{17}, B: 3.4 × 10^{18}, C: 5.0 × 10^{18}, D: 6.7 × 10^{18}. The inserted graph shows the change in the area below the response signals with increasing pulse number. Taken from [40].

(i.e. > 1 min). Also strong diffusion resistance by pore diffusion, as occurs in zeolitic material for example limits the applicability of the method because response signals will be very broad and difficult to integrate.

9.4.2.3 Reversible interaction

Hydrocarbons more often do not interact irreversibly with solid materials. However, due to the interaction with the solid by adsorption and desorption processes, the residence time increases compared to that of an inert gas not interacting with the catalyst (normally Ne, Ar, or N$_2$). In such cases, it is not the pulse intensity but the average response time which is analyzed, e.g. by using the momentum analysis or more complex modeling (see Section 9.3.3). According to the model of Gleaves et al. [3], the activation of desorption can be calculated from the temperature dependence of the residence time using Eqn. 2. Tab. 9.4 is shows a comparison of the heats of adsorption obtained with the TAP method with values obtained by conventional GC pulse techniques [41]. Other data, e.g. for methane on Sm$_2$O$_3$ (5 kJ mol^{-1} [5]) or butane on (VO)$_2$P$_2$O$_7$ catalysts [20] have been reported as well.

Restrictions on the applicability of the mentioned techniques are similar to those for conventional techniques. First, the number of adsorption sites should be large

Tab. 9.4 Enthalpies of reversible desorption for propane and propene from different mixed-metal oxides (T_{ads} = 350 to 450 K); – Comparison of results obtained with the TAP method (C_3H_8:Ne = 1:1; m_{cat} = 0.2–0.3 g; 2–3 × 10^{15} molecules/pulse) and conventional GC pulse method (C_3H_8:Ar:He = 1:1:18; m_{cat} = 10 g; 0.25 μmol/pulse).

Catalyst	Propane $\Delta H/kJ\ mol^{-1}$		Propene $\Delta H/kJ\ mol^{-1}$		Ref.
	TAP	GC pulse	TAP	GC pulse	
Silicalite 1	35.4–36.5				42
α-Bi$_2$MoO$_6$	31.1 ± 4.1	30.2 ± 2.3	50–25*	50.2–8.4	43
Bi-Mo/SiO$_2$		25.9 ± 1.5			44
Ag$_{0.01}$Bi$_{0.85}$V$_{0.54}$Mo$_{0.45}$O$_4$	27.0 ± 1.0	25.3 ± 3.6	28.3 ± 4.7	35.6 ± 5.6	43
Ag$_{0.01}$Bi$_{0.85}$V$_{0.54}$Mo$_{0.45}$O$_4$ /(10 mol%)/γ-Al$_2$O$_3$	27.5 ± 2.0				45
Te$_2$MoO$_x$/Al$_2$O$_3$	32.8 ± 1.1	31.0 ± 2.0	35.8 ± 1.1	40.8 ± 2.0	43
γ-Al$_2$O$_3$	32.4 ± 2.0	31.8			43
α-Sb$_2$O$_4$**					this work
VSb$_1$O$_x$	41.1 ± 1.9		n.d.		this work
VSb$_2$O$_x$	38.5 ± 1.4		n.d.		this work
VSb$_5$O$_x$	48.7 ± 1.1		n.d.		this work
VSb$_5$WO$_x$/Al$_2$O$_3$	40.9 ± 4.2		n.d.		this work

* depending on pulse size; ** no significant interaction; n.d. – not determined due to parallel irreversible interaction.

compared to the incoming pulse size. If the number of sites is small, a dependence of the adsorption enthalpy on the pulse size will be apparent (cf. Tab. 9.4). Second, site heterogeneity on the surface might lead to erroneous values because the contribution of less active sites to the adsorption rates may be underdetermined.

9.4.2.4 Competitive interaction of different gases

Information on competitive adsorption is important for understanding chemical reactions on solid surfaces because the chemical rate mainly depends on the surface concentration of the adsorbed reaction partners. Sequential-pulse experiments can help to understand the underlying mechanisms. The examples in Figs. 9.8(a) and 9.8(b) demonstrate the interaction of propane and propene with a typical ammoxidation catalyst, VSb$_5$O$_x$, in the absence and presence of ammonia on the surface. It can be seen that the response signals of the hydrocarbons are broad for the clean catalyst. Propene interacts more strongly with the solid compared to propane. In the presence of pre-adsorbed ammonia, however, the interaction is dramatically reduced in both cases and the measured response times approach those of a non-interacting inert gas.

From these experiments, different information about the reaction mechanism can be obtained. First, ammonia and the hydrocarbons adsorb on the same sites. Therefore, a reaction between ammonia and hydrocarbon will be slowed down if the surface coverage in ammonia is high. This is indeed observed in experiments performed under atmospheric conditions [43]. Second, the adsorption sites are acidic sites because ammonia is a basic molecule which will preferably interact with Lewis or Brønsted sites [35].

9.4.2.5 Adsorption/desorption properties in the presence of chemical reaction

In case that adsorption properties need to be investigated in the presence of chemical reaction, the above methods fail, because (i) the shape of the response curves is additionally modified due to reaction, or (ii) the irreversibly adsorbed gas is removed from the adsorption site due to reaction, probably leaving a free site behind. Under these circumstances, only complex modeling of all the product pulses will lead to valid information.

9.4.3
Determination of Reaction Mechanisms

By far the most frequent application of the TAP method is the (qualitative) determination of reaction mechanisms by detecting pathways and intermediates which are not accessible under usual reaction conditions. Some examples are listed below.

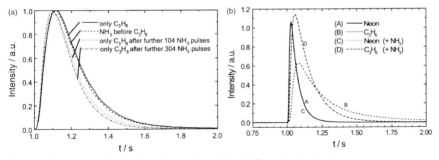

Fig. 9.8 Response signals for pulses of (a) C_3H_8 and (b) C_3H_6 over VSb_5O_x in the presence and absence of adsorbed NH_3 ((a) $C_3H_8/Ne = 1:1$, 3.1 × 10^{14} molecules/pulse; $NH_3/Ne = 1:1$, 2.6 × 10^{16} molecules/pulse; $T = 351$ K; (b) $C_3H_6/Ne = 1:1.2$, $m_{cat} = 150$ mg, $T = 332$ K).

9.4.3.1 **Reaction sequences**

The reaction sequence of an interaction of a hydrocarbon in a sequential manner with a solid surface can easily be determined using the TAP pulse method. Fig. 9.9 shows the responses of exchange products when pure CD_4 was pulsed over Sm_2O_3 [5]. It is evident that the surface reactions of isotopic exchange proceed in a step-wise fashion. It is also obvious that the exchanged methane molecules have longer residence times than CD_4, which passes through the reactor without reaction. This indicates a rather strong dissociative interaction of methane with the surface during the H-D isotopic exchange.

This technique offers a fast means of gaining mechanistic insight into complex reaction networks and has successfully been applied to a number of reactions, such as butane oxidation, propane oxidation, or ammoxidation. For example, Fig. 9.10 shows excerpts of mass scans for a gas flow of a mixture containing pentadiene, O_2, and neon (1:6:1) in a step transient experiment over a $(VO)_2P_2O_7$ catalyst. The changes in the peak intensities clearly indicate that the catalyst properties change in the presence of the gas flow. In the early stages of reaction, phthalic anhydride is formed (e.g. AMU 104, 76), whereas after approximately 10 min its formation is strongly diminished and other products, e.g. 2H-pyran-2-one (AMU 95, 96), are formed. Obviously, reoxidation of the catalyst with the excess gas-phase oxygen does not proceed rapidly enough under the reduced pressure conditions of these flow experiments. Therefore, the redox properties of the VPO catalyst diminish and the intermediate is formed instead of the higher oxidized acid.

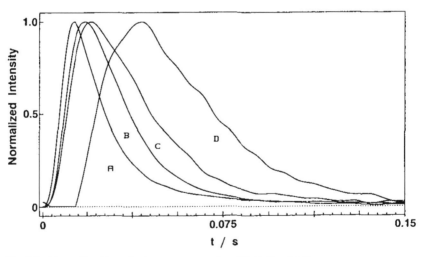

Fig. 9.9 Normalized transient responses for isotopic methanes formed when pure CD_4 was pulsed over Sm_2O_3 at 1073 K (A = CD_4 (enhancement factor $k = 1$),

B = CD_3H ($k = 1.36$),
C = CD_2H_2 ($k = 1.36$),
D = CDH_3 ($k = 1.33$)).

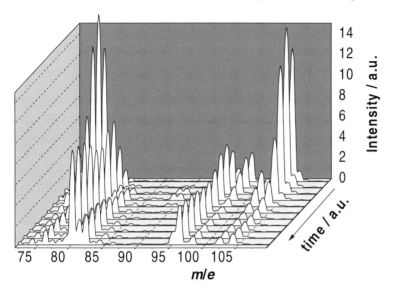

Fig. 9.10 Changes in the mass spectra of the products using a mixture containing 1,3-pentadiene, O_2, and neon as a function of reaction time over a $(VO)_2P_2O_7$ catalyst. $T_R = 823$ K; 1,3-pentadiene/O_2/neon = 1:6:1; $t_R = 60$ min.

The interpretation of these experiments is straighforward for purely sequential reactions of similar components, e.g. isotopes. There are, however, certain cases which could lead to wrong sequences if the functionalities of the products were markedly different, e.g. acidic or basic in character. In general, if a product is less strongly adsorbed than its precursor, this will result in a reverse order of the sequence. Furthermore, due to the fact that the probability of a gas to react increases with increasing residence time in the reactor, the tailing part of a pulse of molecules passing through the reactor is usually diminished by reaction more strongly than the first part. This results in a shift of the maximum of the remaining response signal to shorter residence times, which may consequently lead to misinterpretations. Therefore, the normalizing pulse technique as mentioned above has to be supported by other experiments, usually by pulsing the products themselves on the same catalyst surface using the same reaction parameters.

9.4.3.2 Information about gaseous short-lived intermediates

Gaseous short-lived intermediates of reactions can be directly detected by the mass spectrometer if these are stable enough to leave the reactor. This has been demonstrated for the formation of formyl radicals (CHO·) in the reaction of CO with H_2 over Pd/Al$_2$O$_3$ catalysts [46], and of methyl radicals CH$_3$· in the reaction of methane over rare earth oxides [5]. For the latter species, the temperature depend-

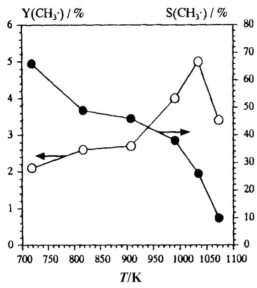

Fig. 9.11 Temperature dependence of the yield and the selectivity of methyl radicals formed in the reaction of methane over Sm_2O_3 (pulse size: 2×10^{15} molecules/pulse; $m_{cat} = 690$ mg). Taken from [40].

ence of selectivity and yield is shown in Fig. 9.11. It was further shown that these radicals can only be observed under the vacuum conditions of the TAP reactor. With increasing pulse size and, hence at increasing pressure, the amount of radicals decreased and coupling products ethane and ethene were formed.

This example already indicates that the special vacuum conditions in the TAP reactor change product composition with respect to atmospheric or pressurized conditions. A similar behavior is observed for the ammoxidation of propane over mixed-metal oxides. Nitrogen is observed under atmospheric conditions as the only direct by-product from ammonia, whereas under TAP conditions its intermediate NO is observed [35]. The TAP method therefore gives new insights into complex reactions which could never be obtained in a usual kinetic experiment.

9.4.3.3 Information about adsorbed short-lived intermediates

In general, the TAP reactor only provides information about the reactants leaving the reactor outlet. For mechanistic considerations, however, the reaction steps taking place on the surface of the catalyst are of great importance. Therefore, modern catalysis often applies *in situ* spectroscopic methods such as DRIFT or Raman spectroscopy to observe surface-bound species directly.

Although such features are not yet possible using the TAP reactor, indirect information on surface-bound species useful for rationalizing complex mechanistic features can be obtained by the sequential-pulse technique. As already explained, following introduction of the first pulse the surface is populated with a certain amount of adsorbed species, which then can react with a component subsequently introduced via the second pulse valve.

The pulsed amount of gas over the catalyst and the strength of adsorption mainly determine the surface coverage of the adsorbed species. These parameters can indirectly be controlled by the pulse size of the first pulse and the time difference between the first and second pulses. With increasing pulse size, the amount of gas interacting with the catalyst increases, whereas with increasing time difference the amount of adsorbate diminishes due to desorption processes or reaction with the catalyst itself. For our example of propane ammoxidation it has been discussed whether strongly bound Me=NH groups or loosely bound $NH_{3,ads}$ or NH_4^+ participate in the nitrile formation.

A sequential-pulse experiment in which first NH_3 was pulsed over the V/Sb/O catalyst followed, after a time lag Δt, by C_3H_8 could answer the question. Fig. 9.12 shows the change in the acrylonitrile signal in response to the second C_3H_8 pulses at various time lags Δt. With increasing time difference, the amount of acrylonitrile formed decreases. This indicates that the NH_x intermediate formed by interaction of NH_3 with the oxide is short-lived and the half-life time can be calculated from the decrease in acrylonitrile formation. As expected, the lifetime was found to depend on the type of catalytic material used (see Tab. 9.5).

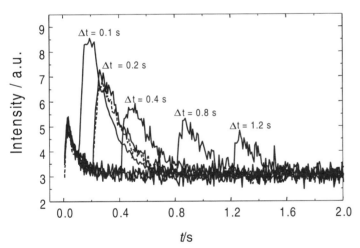

Fig. 9.12 Sequential pulsing of 1. NH_3 and 2. C_3H_8 over VSb_5O_x; dependence of the amount of acrylonitrile formed on the time difference Δt between the two pulses (m_{cat} = 250 mg; T_R = 723 K; NH_3/He = 1:3; 4×10^{16} molecules/pulse; C_3H_8/Ne = 1:9; 2×10^{16} molecules/pulse).

Catalyst	Half-life time [s]
VSb_5O_x	0.5
VSb_5O_x/Al_2O_3	10
$(VO)_2P_2O_7$	>80

Tab. 9.5 Half-life time of NH_x species formed by interaction of NH_3 with oxide catalysts (T_R = 723 K; NH_3/He = 1:3; 4×10^{16} molecules/pulse).

9.4.3.4 Identification of different active sites and properties of the reacting solid

We have already mentioned that isotopically labeled compounds are favorable for TAP investigations, mainly because they can be used to discriminate different possible reaction paths. Isotopically marked compounds usually have similar kinetics (except hydrogen, because of its mass difference compared to deuterium [47, 48]). Furthermore, in most cases, they can easily be distinguished by mass spectrometry and due to the special conditions in the TAP reactor only small amounts are necessary, which significantly reduces the costs of such experiments. ^{18}O, ^{13}C, ^{15}N, and ^2H have successfully been used to investigate reaction mechanisms and pathways of organic molecules. However, they can also be used to obtain information on the surface, subsurface, and bulk of the solid catalysts. This is demonstrated in the following example.

Propene reacts on V/Sb/O catalysts forming acrolein, CO, and CO_2 under the vacuum conditions in the TAP reactor. The oxygen species that determines the reaction is nucleophilic lattice oxygen and thus the reaction proceeds in according to a Mars–van Krevelen type redox mechanism [22]. Due to the heterogeneity of the catalyst [23], two different catalytic active sites exist on the solid surface. Their different performance in the catalytic reaction can be demonstrated using isotopic step transient experiments (Fig. 9.13). In this type of experiment, a constant gas flow containing propene and labeled oxygen ($^{18}O_2$) is passed over the catalyst ($T_R = 670$ K; $p_{tot} < 1$ Torr). The gas composition is measured at the reactor outlet and the isotopic distribution of the oxygen-containing products (acrolein, CO, and CO_2) is calculated. After 1 h, the flow is stopped and the system remains in vacuo. The temperature is increased by 100 K for one additional hour. After a decrease in temperature to 670 K, the gas flow containing $^{18}O_2$ is started again.

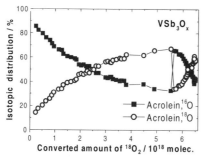
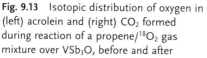
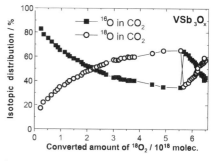

Fig. 9.13 Isotopic distribution of oxygen in (left) acrolein and (right) CO_2 formed during reaction of a propene/$^{18}O_2$ gas mixture over VSb$_3$O$_x$ before and after heating in vacuum to 770 K without gas flow ($m_{cat} = 150$ mg; $T_R = 670$ K; $Q = 0.57$ mL/min).

At the very beginning of the propene/oxygen gas flow, only acrolein and CO_2 containing exclusively ^{16}O are observed. This indicates that only lattice oxygen is involved in the reaction. With increasing reaction time, i.e. with increasing amounts of converted $^{18}O_2$ from the gas phase involved in reoxidizing the reduced metal oxide sites, the proportion of molecules containing ^{18}O increases. After 1 h (i.e. 5.5×10^{18} molecules of converted $^{18}O_2$), approximately 65% of the oxygen in the products acrolein and CO_2 is ^{18}O. However, after 1 h under vacuum at elevated temperature the ^{18}O content is strongly reduced, indicating that under these conditions surface oxygen can diffuse into the catalyst bulk and exchange with lattice ^{16}O oxygen atoms. It is interesting to note that for both products the same isotopic distribution is observed. For the third product CO, however, a completely different behavior is observed (Fig. 9.14).

Even at the very beginning of the reaction, the ^{18}O labeled fraction of CO exceeds 80%. Furthermore, no significant change during the vacuum heating is observed. This behavior can be explained in terms of a low number of highly reactive sites which rapidly exchange oxygen by reaction with gaseous oxygen. However, there is no interrelation between these sites and the bulk of the catalyst, because no change in the CO isotopic composition is observed after temperature increase under non-reactive conditions.

Furthermore, these results reveal that two different sites exist on V/Sb/O catalysts, of which one converts propene mainly into acrolein and CO_2 whereas the second mainly produces CO.

9.5
Concluding Remarks

The investigation of the kinetics of complex gas-phase catalytic reactions and the determination of the underlying kinetic parameters has been and still is one of the main tasks for the application of the TAP reactor. The simple hydrodynamics

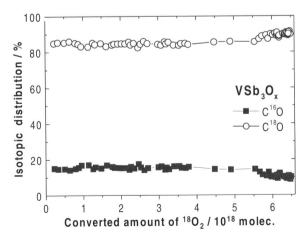

Fig. 9.14 Isotopic distribution of oxygen in CO formed during reaction of a propene/$^{18}O_2$ gas mixture over VSb$_3$O$_x$ before and after heating in vacuum to 770 K without gas flow (m_{cat} = 150 mg; T_R = 670; Q = 0.57 mL/min).

(mostly Knudsen diffusion) allows one to focus on the intrinsic kinetic equations. Consequently, kinetic parameters have been successfully determined for several reactions. Furthermore, the special conditions often allow evaluation of the kinetics on the basis of elementary steps. However, it should always be taken into account that the vacuum conditions applied in the TAP reactor strongly differ from the "real" technical conditions (the so-called *pressure gap*). This results in major drawbacks, which can hinder the bridging of this gap. The user of the TAP method should therefore be aware of the following problems that may occur during extrapolation of the results:

1. The surface of the catalyst used in the TAP reactor is a clean surface, mostly free from adsorbed reactants at the beginning of an experiment.
2. The catalyst surface in vacuo may differ from that under pressurized conditions (e.g. surface reconstruction).
3. The effects observed may not be an average result from all active sites, but sites that exhibit a higher reactivity might be over-represented.
4. Sometimes different intermediates and products are observed compared to studies performed at atmospheric or pressurized conditions and the pressure dependence of their formation is only rarely understood. This may lead to a different mechanistic model.

Mechanistic TAP reactor studies have allowed the determination of physically meaningful kinetic parameters for elementary step rate constants comparable to those obtained in experiments performed under technical conditions. Nevertheless, the measurement of steady-state kinetics near industrial conditions is supposed to provide a safer basis for extrapolation beyond the measured data in terms of reaction rates.

The presented results show that the unique variety of methods available in the TAP reactor system allows a systematic investigation of complex reaction mechanisms and pathways of heterogeneously catalyzed gas-phase reactions. A full elucidation of a chemical reaction by TAP experiments alone is, however, not possible. The authors would like to recommend the users of the TAP method to support their investigations with well-established surface and bulk characterization techniques to cope with the complexity of the solids they are working with in order to avoid misinterpretations.

References

1 H. Kobayashi, M. Kobayashi, "Transient Response Method in Heterogeneous Catalysis", *Catal. Rev.* **10** (1974) 139–176.
2 C. O. Bennett, "The Transient Method and Elementary Steps in Heterogeneous Catalysis", *Catal. Rev.* **13** (1976) 121–148.
3 J. T. Gleaves, J. R. Ebner, T. C. Kuechler, "Temporal Analysis of Products (TAP) – A Unique Catalyst Evaluation System with Submillisecond Time Resolution", *Catal. Rev.-Sci. Eng.* **30**(1) (1988) 49–116.

4 J. R. Ebner, J. T. Gleaves (Monsanto, St. Louis, MO), "Method and Apparatus for Carrying out Catalyzed Chemical Reactions and for Studying Catalysts", *U.S. Patent no. 4,626,412*, Dec. 2, 1986.

5 O. V. Buyevskaya, M. Rothaemel, H. W. Zanthoff, M. Baerns, "Transient Studies on Reaction Steps in the Oxidative Coupling of Methane over Catalytic Surfaces of MgO and Sm$_2$O$_3$", *J. Catal.* **146** (1994) 346–357.

6 O. V. Buyevskaya, M. Rothaemel, H. W. Zanthoff, M. Baerns, "Transient Studies on the Role of Oxygen Activation in the Oxidative Coupling of Methane over Sm$_2$O$_3$, Sm$_2$O$_3$/MgO, and MgO Catalytic Surfaces", *J. Catal.* **150** (1994) 71–80.

7 J. R. Ebner, J. T. Gleaves, T. C. Kuechler, T. P. Li, "Studies of Binary Oxide Catalysts for the Ammoxidation of Methanol to Hydrogen Cyanide", *Symposium on Chemicals from Syngas and Methanol Presented before the Division of Petroleum Chemistry, Inc. American Chemical Society, New York Meeting, 1986.*

8 J. R. Ebner, J. T. Gleaves (Monsanto Company, St. Louis, MO), "Apparatus for Carrying out Catalyzed Chemical Reactions and for Studying Catalysis", *U.S. Patent no. 5,009,849*, Apr. 23, 1991.

9 J. R. Ebner, J. T. Gleaves (Monsanto Company, St. Louis, MO), "Method and Apparatus for Carrying out Catalysed Chemical Reactions and for Studying Catalysis", *U.S. Pat. no. 5,264,183*, Nov. 23, 1993.

10 G. D. Svoboda, J. T. Gleaves, P. L. Mills, "New Method for Studying the Pyrolysis of VPE/CVD Precursors under Vacuum Conditions. Application to Trimethylantimony and Tetramethyltin", *Ind. Eng. Chem. Res.* **31** (1992) 19–29.

11 D. S. Lafyatis, G. Creten, G. F. Froment, "TAP Reactor Study of the Partial Oxidation of Methanol to Formaldehyde Using an Industrial Fe/Cr/Mo Oxide Catalyst", *Appl. Catal. A: General* **120** (1994) 85–103.

12 A. C. van Veen, H. Zanthoff, O. Hinrichsen, M. Muhler, "A Fixed-Bed Microreactor for Transient Kinetic Experiments with Strongly Adsorbing Gases under High Vacuum Conditions", *J. Vac. Sci. Technol. A,* **19** (2001) 651–655.

13 J. T. Gleaves, P. T. Harkins, "Apparatus for Catalyst Analysis", *U.S. Patent no. 5,039,489*, Aug. 13, 1991.

14 D. R. Coulson, P. L. Mills, K. Kourtakis, J. J. Lerou, L. E. Manzer, "Kinetics of the Reoxidation of Propylene-Reduced γ-Bismuth Molybdate: A TAP Reactor Study", *Stud. Surf. Sci. Catal.* **72** (1992) 305–316.

15 G. Creten, D. S. Lafyatis, G. F. Froment, "Transient Kinetics from the TAP Reactor System. Application to the Oxidation of Propylene to Acrolein", *J. Catal.* **154** (1995) 151–162.

16 J. T. Gleaves, "Apparatus for Study and Analysis of Products of Catalytic Reaction", *U.S. Patent no. 5,376,335*, Dec. 27, 1994.

17 S. C. van der Linde, T. A. Nijhuis, F. H. M. Dekker, F. Kapteijn, J. A. Moulijn, "Mathematical Treatment of Transient Kinetic Data: Combination of Parameter Estimation with Solving the Related Partial Differential Equations", *Appl. Catal. A: General* **151** (1997) 27–57.

18 Y. Schuurman, A. Pantazidis, C. Mirodatos, "The TAP-2 Reactor as an Alternative Tool for Investigating FCC Catalysts", *Chem. Eng. Sci.* **54** (1999) 3619–3625.

19 T. A. Nijhuis, L. J. P. van den Broeke, M. J. G. Linders, J. M. van de Graaf, F. Kapteijn, M. Makkee, J. A. Moulijn, "Measurement and Modeling of the Transient Adsorption, Desorption and Diffusion Processes in Microporous Materials", *Chem. Eng. Sci.* **54** (1999) 4423–4436.

20 B. Kubias, U. Rodemerck, H.-W. Zanthoff, M. Meisel, "The Reaction Network of the Selective Oxidation of *n*-Butane on $(VO)_2P_2O_7$ Catalysts: Nature of Oxygen–Containing Intermediates", *Catal. Today* **32** (1996) 243–253.

21 G. Creten, F.-D. Kopinke, G. F. Froment, "Investigation of the Oxidation of *o*-Xylene over a Vanadia/Titania Catalyst by Means of the TAP Reactor", *Can. J. Chem. Eng.* **75** (1997) 882–891.

22 H. W. Zanthoff, S. A. Buchholz, A. Pantazidis, C. Mirodatos, "Selective and Nonselective Oxygen Species Determining the Product Selectivity in the Oxidative Conversion of Propane over Vanadium Mixed Oxide Catalysts", *Chem. Eng. Sci.* **54** (1999) 4397–4405.

23 H. W. Zanthoff, W. Grünert, S. Buchholz, M. Heber, L. Stievano, F. E. Wagner, G. U. Wolf, "Bulk and Surface Structure and Composition of V/Sb Mixed-Oxide Catalysts for the Ammoxidation of Propane", *J. Molec. Catal. A: Chemical* **162** (2000) 443–462.

24 O. Dewaele, G. F. Froment, "TAP Study of the Sorption of CO and CO_2 on γ-Al_2O_3", *Appl. Catal. A: General* **185** (1999) 203–210.

25 N. C. Rigas, G. D. Svoboda, J. T. Gleaves, in: *Catalytic Selective Oxidation* (Eds.: S. T. Oyama, J. W. Hightower), ACS Symposium Series, Washington DC, 1993, p. 183–203.

26 B. S. Zou, M. P. Duducovic, P. L. Mills, "Modeling of Pulsed Gas Transport Effects in the TAP Reactor System", *J. Catal.* **145** (1994) 683–696.

27 J. T. Gleaves, G. S. Yablonskii, P. Phanawadee, Y. Schuurman, "TAP-2: An Interrogative Kinetics Approach", *Appl. Catal. A* **160** (1997) 55–88.

28 C. Mirodatos, "How Transient Kinetics May Unravel Methane Activation Mechanisms", *Stud. Surf. Sci. Catal.* **119** (1998) 99–106.

29 M. Rothaemel, B. Baerns, "Modeling and Simulation of Transient Adsorption and Reaction in Vacuum Using Temporal Analysis of Products Reactor", *Ind. Eng. Chem. Res.* **35** (1996) 1556–1565.

30 M. Soick, D. Wolf, M. Baerns, "Determination of Kinetic Parameters for Complex Heterogeneous Catalytic Reactions by Numerical Evaluation of TAP Experiments", *Chem. Eng. Sci.* **55** (2000) 2875–2882.

31 S. O. Shekhtman, G. S. Yablonsky, S. Chen, J. T. Gleaves, "Thin-Zone TAP Reactor–Theory and Application", *Chem. Eng. Sci.* **54** (1999) 4371–4378.

32 G. S. Yablonskii, I. N. Katz, P. Phanawadee, J. T. Gleaves, "Symmetrical Cylindrical Model for TAP Pulse Response Experiments and Validity of the One-Dimensional TAP Model", *Ind. Eng. Chem. Res.* **36** (1997) 3149–3153.

33 M. Soick, O. Buyevskaya, M. Höhenberger, D. Wolf, "Partial Oxidation of Methane to Synthesis Gas over Pt/MgO: Kinetics of Surface Processes", *Catal. Today* **32** (1996) 163–169.

34 A. Pantazidis, S. A. Buchholz, H. W. Zanthoff, Y. Schuurman, C. Mirodatos, "A TAP Reactor Investigation of the Oxidative Dehydrogenation of Propane over a V/Mg/O Catalyst", *Catal. Today* **40** (1998) 207–214.

35 H. W. Zanthoff, S. A. Buchholz, O. Y. Ovsitser, "Ammoxidation of Propane to Acrylonitrile on V/Sb/O Catalysts. Role of Ammonia in the Reaction Pathways", *Catal. Today* **32** (1996) 291–296.

36 J. H. B. J. Hoebink, J. P. Huinink, G. B. Marin, "A Quantitative Analysis of Transient Kinetic Experiments: The Oxidation of CO by O_2 over Pt", *Appl. Catal. A: General* **160** (1997) 139–151.

37 F. Konietzni, H. W. Zanthoff, W. F. Maier, "The Role of Active Oxygen in the AMM-V$_x$Si-Catalysed Selective Oxidation of Toluene", *J. Catal.* **188** (1999) 154–164.

38 M. Fathi, F. Monnet, Y. Schuurman, A. Holmen, C. Mirodatos, "Reactive Oxygen Species on Platinum Gauzes during Partial Oxidation of Methane into Synthesis Gas", *J. Catal.* **190** (2000) 439–445.

39 W. L. M. Weerts, M. H. J. M. de Croon, G. B. Marin, "The Adsorption of Silane Disilane and Trisilane on Polycrystalline Silicon: A Transient Kinetic Study", *Surf. Sci.* **367** (1996) 321–339.

40 M. Rothaemel, *Transientenexperimente zur Aufklärung von Adsorptionsvorgängen und katalytischen Reaktionen an Festkörperoberflächen: Modellierung, Simulation und Experiment* (Ph. D. thesis, Ruhr-Universität Bochum, 1995).

41 R. L. Laub, R. L. Pescok, *Physicochemical Application of Gas Chromatography* (John Wiley & Sons, New York, 1978).

42 T. A. Nijhuis, *Towards a New Propene Epoxidation Process – Transient Adsorption and Kinetics Measurements Applied in Catalysis*, (Ph. D. thesis, TU Delft, 1998).

43 H. W. Zanthoff, *Mechanismen und Wirkungsweisen heterogener Katalysatoren in der oxidativen Funktionalisierung von Alkenen – Die Ammoxidation von Propan zu Acrylnitril* (Habilitation thesis, Ruhr-Universität Bochum, 1999).

44 J. Strnad, M. Krivanek, "Chromatographic Determination of Adsorption Heats of Propylene, Propane, Acrolein, and CO_2 on Oxides of Bi, Mo, Bi-Mo, Bi-Mo/SiO$_2$, and on SiO$_2$", *J. Catal.* **23** (1971) 253–258.

45 O. Seel, *Grundlegende Untersuchungen zur Katalyse der Partiellen Oxidation von Propan zu Acrolein* (Ph. D. thesis, Ruhr-Universität Bochum, 1995).

46 M. Rothaemel, H. W. Zanthoff, M. Baerns, "Formation of CHO During Interaction of CO and H_2 on Alumina-Supported Pd Catalysts", *Catal. Lett.* **28** (1994) 321–328.

47 N. W. Cant, W. K. Hall, "Catalytic Oxidation VI. Oxidation of Labeled Olefins over Silver", *J. Catal.* **52** (1978) 81–94.

48 R. A. van Santen, J. Moolhuysen, W. M. H. Sachtler, "The Role of CH Bond Breaking in the Epoxidation of Ethylene", *J. Catal.* **65** (1980) 478–480.

Subject Index